电池科普与环保①

Battery Science Popularization and Environmental Protection①

电池大家族

The Battery Family

（中英对照版）
(Chinese-English Version)

马建民 / 主编　　*Edited by: Ma Jianmin*
咪柯文化 / 绘　　*Illustrated by: Micco Culture*
何　敏 / 译　　　*Translated by: He Min*

电子科技大学出版社
University of Electronic Science and Technology of China Press
·成都·

前言 Preface

人类对能源的探索永不停止

人类对能源的探索，从来就未曾停止。由于地球上可供开采的煤炭、石油、天然气等非再生能源十分有限，因此，现在全世界都将目光聚焦于太阳能、风能、核能、潮汐能等再生能源的开发与利用。

能源问题是关系国家安全、社会稳定和经济发展的重大战略问题。优化资源配置，提高能源的有效利用率，对于人类的生存和国家的发展都具有十分重要的意义。

如何积极发展新能源是人类必须共同面对的一项重大技术课题。新能源技术的不断进步，特别是动力系统的不断改进，为能源结构的转型提供了可能。然而，虽然新能源的类型很多，但世界上至今还没有实用的、经济有效的、大规模的直接储能方式。因此，人类还不得不借助间接的储能方式。

电能，作为支撑人类现代文明的二次能源，既能满足大量生产、集中管理、自动化控制和远距离输送的需求，又具有使用方便、洁净环保、经济高效的特点。因此，电能可以替代其他能源，提高能源的利用效率。

我们今天所有的可移动电子设备，其运行都离不开电池。电池的出现使人类的生活更加便捷，特别是在信息时代来临之后，电池的重要性更为突出。我国不仅是世界排名第一的电池生产大国，也是世界排名第一的电池消费大国。

人类虽然在电池的研究方面已经取得了丰硕的成果，但还一直在寻找更好的电能储存介质。随着科学的发展、新能源技术的成熟，在未来，哪一种类型的电池能够脱颖而出还未可知。希望此书能激发孩子们对电池的兴趣，让他们在未来为我们揭晓谜底。

马建民

2024 年 3 月

Endless Exploration of Energy

Since ancient times, humans' quest for energy has never ceased. Given the limited reserves of non-renewable energy like coal, oil, and natural gas on Earth, the use of renewable energy like solar, wind, nuclear, and tidal power, has become the new global focus.

Energy is a major strategic issue that bears on national security, social stability and economic development. How to allocate and use energy in a better way means a lot for both individuals and countries.

How to develop new energy is a major technological topic facing humanity. With the development of new energy technology, especially the power system, energy structure transformation is made possible. However, despite various kinds of new energy, there is yet to be a practical, cost-effective, large-scale way of direct energy storage. Therefore, we have to resort to indirect methods to store energy.

Electricity, as a secondary energy propelling modern civilization, can support mass production, centralized management, automated control, and long-distance transmission. At the same time, electricity is clean, economical, efficient, environmentally friendly and easy to use. We can replace other energy with electricity to use energy better.

All the electronic mobile devices today can't operate without batteries. Batteries make our life more convenient. Its importance grows even more prominent with the advent of the Information Age. China is now the world's biggest producer and consumer of batteries.

Although we have gained so much in battery research, researchers are still looking for a better medium for power storage. With progress in science and new energy technologies, which type of battery will stand out still awaits our exploration. Hopefully, this book will spur children's interest in batteries and one day make them tell us the answer in the future.

Ma Jianmin

March 2024

电池大家族·族谱
Family Tree of Battery

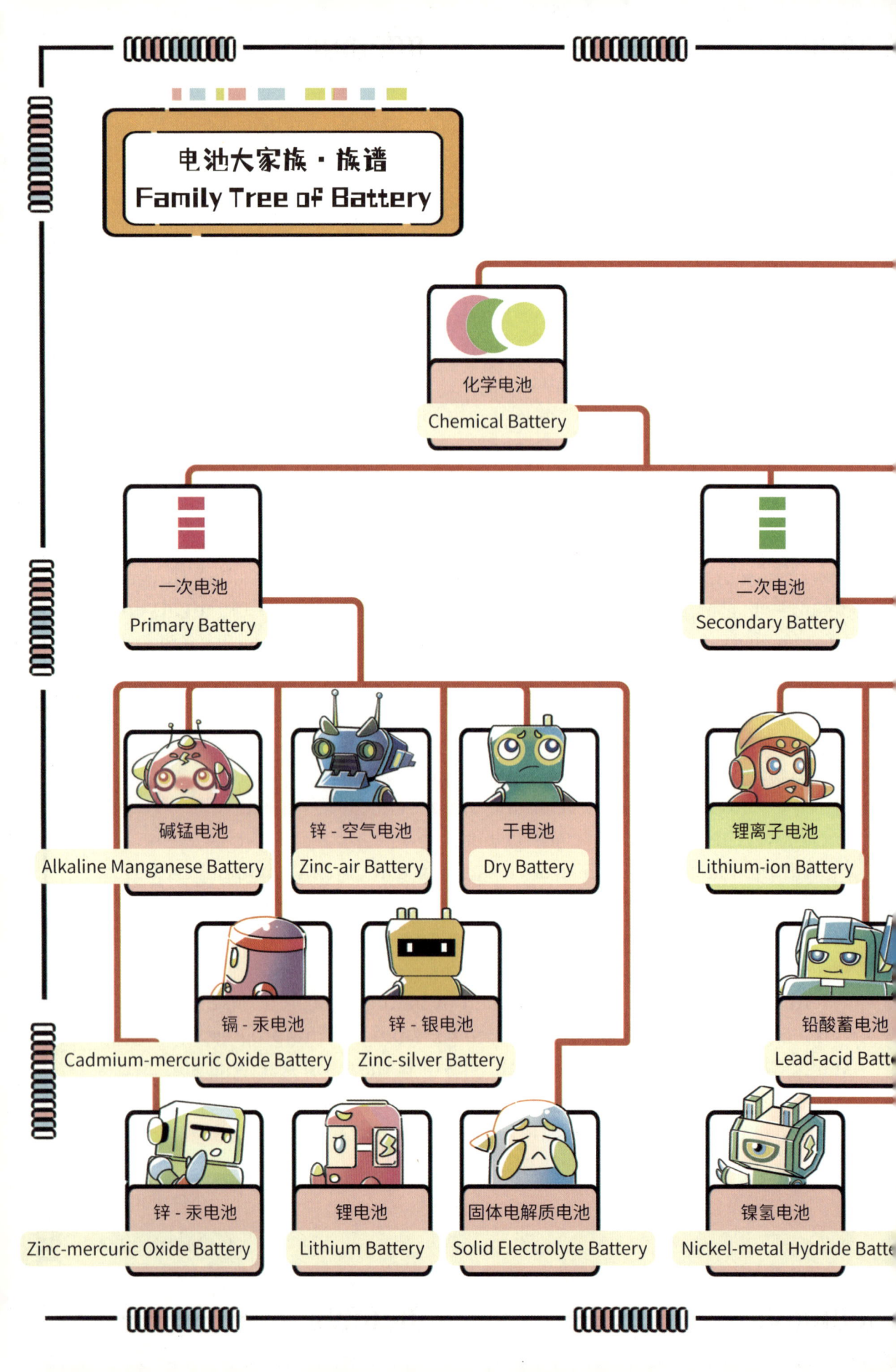

- 化学电池 / Chemical Battery
  - 一次电池 / Primary Battery
    - 碱锰电池 / Alkaline Manganese Battery
    - 锌-空气电池 / Zinc-air Battery
    - 干电池 / Dry Battery
    - 镉-汞电池 / Cadmium-mercuric Oxide Battery
    - 锌-银电池 / Zinc-silver Battery
    - 锌-汞电池 / Zinc-mercuric Oxide Battery
    - 锂电池 / Lithium Battery
    - 固体电解质电池 / Solid Electrolyte Battery
  - 二次电池 / Secondary Battery
    - 锂离子电池 / Lithium-ion Battery
    - 铅酸蓄电池 / Lead-acid Battery
    - 镍氢电池 / Nickel-metal Hydride Battery

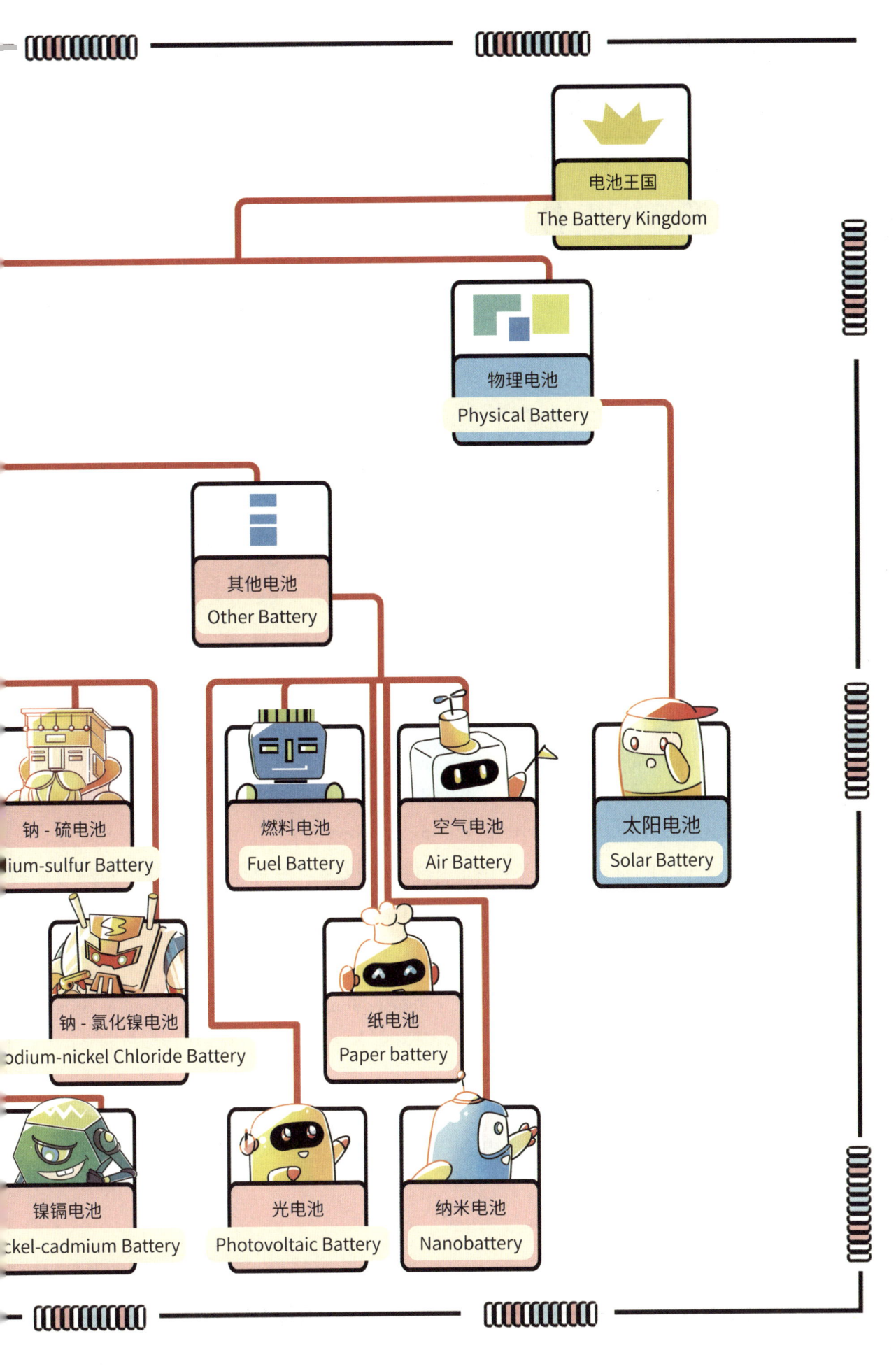

电池王国
The Battery Kingdom

物理电池
Physical Battery

其他电池
Other Battery

- 钠-硫电池 / Sodium-sulfur Battery
- 燃料电池 / Fuel Battery
- 空气电池 / Air Battery
- 太阳电池 / Solar Battery
- 钠-氯化镍电池 / Sodium-nickel Chloride Battery
- 纸电池 / Paper battery
- 镍镉电池 / Nickel-cadmium Battery
- 光电池 / Photovoltaic Battery
- 纳米电池 / Nanobattery

故事导读 Introduction

电池王国是一个庞大的国度，其中生活着许许多多的电池家族，每个家族的电池人都有着独特的本领。他们勤劳能干，驱动各种设备运转，促进人类世界不断发展。

在电池王国里，每天都有故事发生。这不，在电池王国成立纪念日这天，一条消息席卷了全国——五年一届的电池大赛启动了。各种类型的电池都摩拳擦掌，准备大展身手。

经过预选赛和初赛，两个来自不同家族的电池人脱颖而出。赛事正在如火如荼地进行着，国王突然宣布了一个爆炸性的消息——冠军得主将是未来的国王。大家有点不敢相信，国王竟会主动放弃王权？

大家议论纷纷，原来，这背后有着许多不为人知的故事……让我们一起去看看，这场电池王国里的较量吧！

The Battery Kingdom is a vast land with many families. Each family has its own unique abilities. They work hard, make machines run, and help the human world keep moving forward.

In this Kingdom, there are things going on every day. On the anniversary of the kingdom's founding, a piece of news spread everywhere—the National Battery Contest, held once every five years, kicked off! Batteries from all kinds of families were getting ready to show off their skills.

After the tryouts and the first round, two batteries from different families made it to the final. The contest was heating up, when suddenly, the King announced exciting news—the winner would become the next King! It was hard to believe that the King would ever give up the to be a king.

Everybody was talking about it, and soon, people realized there was more to this story than they knew… Let's see what happened in this big contest in the Battery Kingdom!

角色介绍
Characters

锂锂 Lithium Li

家族：锂离子电池
Family: Lithium-ion Battery

氢天天 Hydrogen Hero

家族：镍氢电池
Family: Nickel-metal Hydride Battery

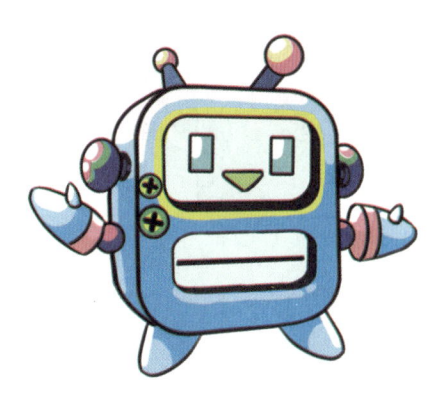

机器人 X Robot X

大铅　Big Lead

家族：铅酸蓄电池

Family: Lead-acid Battery

锌博士　Dr. Zinc

家族：锌-空气电池

Family: Zinc-air Battery

镍霸　Tyrant Nickel

家族：镍镉电池

Family: Nickel-cadmium Battery

王一硫　King Monosulfide

家族：钠－硫电池

Family: Sodium-sulfur Battery

王一氯　King Monochlorine

家族：钠－氯化镍电池

Family: Sodium-sulfur Battery

锰叨叨　Mnag

家族：碱锰电池

Family: Alkaline-manganese Battery

目 录
Table of Contents

1 相约决赛
See You in the Final — /001

2 初赛——电池三项大比拼
Preliminary Round—The Big Battery Triathlon — /013

3 极具挑战的晋级赛
Challenging Advancement Race — /037

4 艰难晋级！我有绝技
Tough Advancement! I Have a Secret Skill — /063

5 新王诞生
Rise of the New King — /077

电池大揭秘
Secrets behind Batteries

/093

电池的定义
Definition of Batteryies ···094

电池的分类
Types of Batteries ···099

探秘锂离子电池
Exploring Lithium-ion Batteries ···103

探秘铅酸蓄电池
Exploring Lead-acid Batteries ···108

探秘碱锰电池
Exploring Alkaline-manganese Batteries ·································111

探秘镍氢电池
Exploring Nickel-metal Hydride Batteries ································115

探秘镍镉电池
Exploring Nickel-cadmium Batteries ·····································118

探秘镍镉电池与镍氢电池
Exploring Nickel-cadmium Batteries and Nickel-metal Hydride Batteries
···121

探秘锌 - 空气电池

Exploring Zinc-air Batteries……………………………………123

探秘钠 - 氯化镍电池

Exploring Sodium-nickel Chloride Batteries……………………125

探秘钠 - 硫电池

Exploring Sodium-sulfur Batteries………………………………128

1

相约决赛

See You in the Final

一大早，机器人 X 清脆的声音打断了锂锂的工作。啪！电视机被打开了，原来是电池王国五年一度的"电池竞技大赛"即将举行。

Early in the morning, the clear voice of Robot X interrupted Lithium Li's work. Pop! The TV was turned on, and it turned out that the once-every-five-year "Battery Competition" of the Battery Kingdom was about to be held.

滴——滴——
惊爆消息，锂锂快看！

Beep— Beep—
Breaking news! Lithium Li, look! look!

"是的，这场赛事意义重大，所以我会亲自嘉奖决赛的冠军……"

"Yes, this event is really important, so I'll give the prize to the winner in person."

呀，全国各个电池家族的人都来了！

Wow, members from all battery families across the country are here!

姓名：	锂锂
种族：	二次电池
家族：	锂离子电池
适应温度：	−20 ~ 60 ℃

主要应用领域：新能源汽车、航空航天、电动工具、电动玩具等。

Name:	Lithium Li
Type:	Secondary Battery
Family:	Lithium-ion Battery
Operating temperature:	−20~60 ℃

Main application areas: new energy vehicles, aerospace, electric power tools, electric toys, etc.

全国电池竞技大赛冲冲冲

锂锂赶紧来到了预选赛场入口，熙熙攘攘的人群已经把报名处围了个水泄不通。

Lithium Li hurried to the preliminary round venue. A crowd has surrounded the registration area.

我先来的，你靠后！
I came first. You step back!

你才早来1秒钟！
You're only one second earlier!

别挤，别挤，大家都有机会上场。
Don't push! Everyone gets a chance.

嘻嘻，我已经报完名了！
Hehe, I've signed up!

赛场内，各式各样的机器人停在各自的工作仓，令人目不暇接。

In the arena, robots of all sorts were stationed in their own workplace. Dazzling!

火热进行中!!

预选赛规则　Preliminary Round Rules

参赛选手自选并驱动某一类型机器人，完成的动作越多、越精准，得分越高，满分 10 分。

Contestants choose and drive a certain type of robot by themselves. The more precise actions, the higher scores. Full score is 10.

耶，我终于可以大展风采了，国家级赛事啊，为家族争光，想想就激动！

Yeah! It's a national competition. I can finally prove myself and bring glory to my whole family. So excited!

电池竞技大赛预选赛正式开始，锂锂进入8号工作仓，这是制作饮料的机器人的工作仓。锂锂娴熟的操作取得了高达9分的优异成绩。

The preliminary round of the Battery Competition officially began. Lithium Li entered workstation No.8, which was the workspace for beverage-making robot. Lithium Li's skilled operation earned an outstanding score of 9 points.

选手进入工作仓。
The contestant entered the workstation.

比赛开始
The competition begins

比赛结束
The competition ends

锂锂非常高兴,哼着小曲儿走出工作仓。

在经过大铅的工作仓门口时,锂锂用余光扫到,他的成绩也是9分,不由对这位胖胖的、萌萌的,还一副学识渊博模样的电池人产生了兴趣。

Lithium Li was very happy, humming a tune as he walked out of the workstation.

As Lithium Li walked past Big Lead's workstation, he saw another 9-point winner. He got interested in the chubby, cute, and smart-looking battery.

姓名:	大铅
种族:	二次电池
家族:	铅酸蓄电池
适应温度:	−20～60 ℃
主要应用领域:	交通工具、电力系统、通信设备、工业设备、国防军工等。

Name:	Big Lead
Type:	Secondary Battery
Family:	Lead-acid Battery
Operating temperature:	−20~60 ℃
Main application areas:	vehicles, power systems, communication equipment, industrial equipment, and national defense industries, etc.

和我一样能得9分,交个朋友如何?

You got a 9 just like me. Want to be friends?

英雄所见略同!

Great minds think alike!

正当他们沟通交流之时，10号工作仓内，氢天天驾驶电动货车以极快的速度完成了配货。

While they were talking, Hydrogen Hero in workstation No.10 quickly completed the delivery with an electric cart.

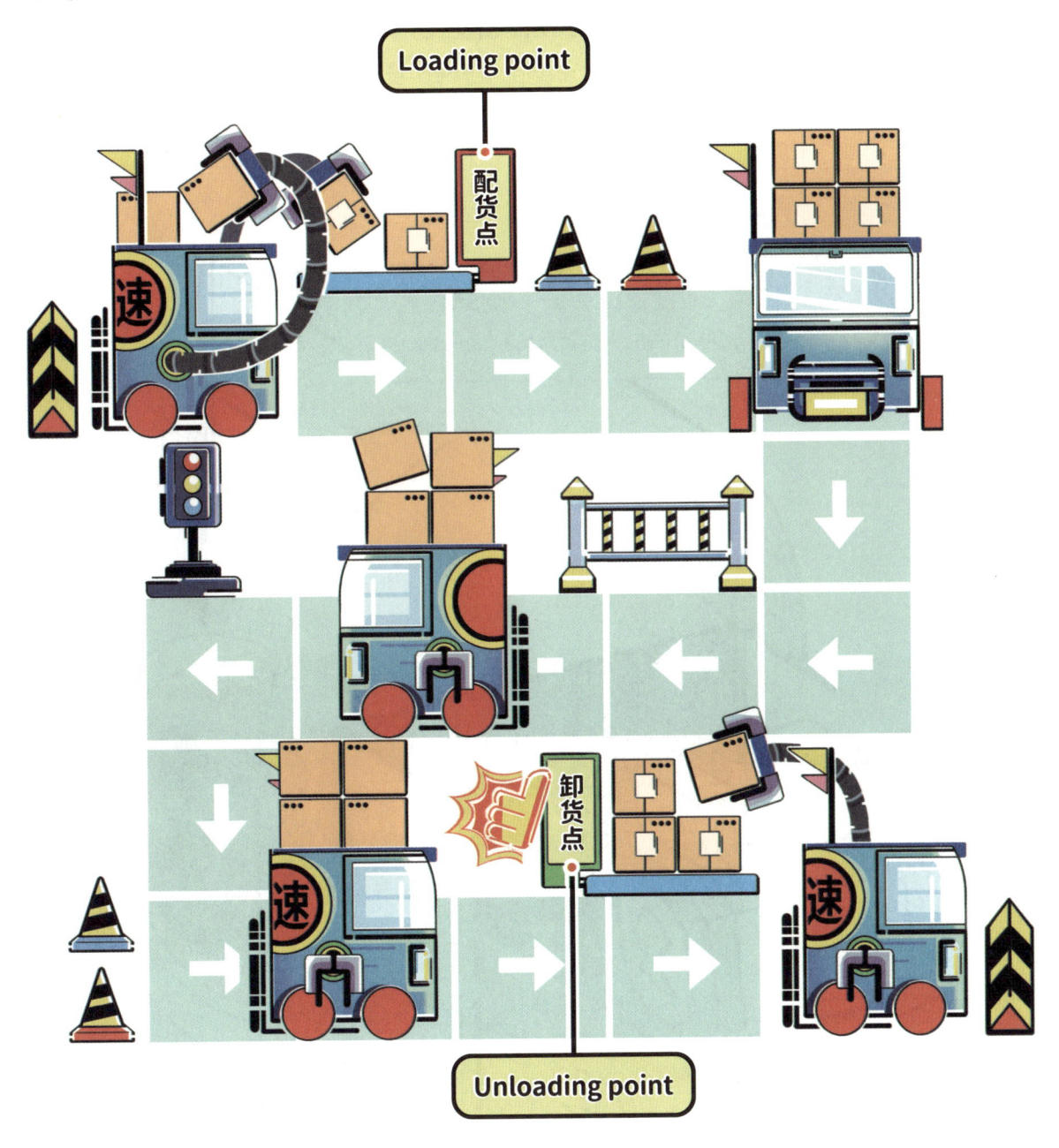

随后清点、装车、加固、发车、到站、卸货，动作一气呵成，数量分毫不差，货物完好无损。

Counting, loading, securing, sending, arriving, and unloading were done smoothly. Exact in number! Safe and sound!

氢天天同样取得了9分的高分，观众台上爆发的掌声如潮水般此起彼伏，这也引起了锂锂和大铅的注意。他们正走到赛场门口，回头看到这一幕，不约而同地发出了赞叹。

Hydrogen Hero also earned a high score of 9 points, and the audience's applause came in waves, catching the attention of Lithium Li and Big Lead. They were walking to the entrance of the arena, turned around and saw this scene, and exclaimed in admiration at the same time.

> 我们又有一个9分得主诞生了，
> Another 9-point winner—

> 他是10号工作仓位选手——氢天天！
> contestant in workstation No.10, Hydrogen Hero!

姓名：	氢天天
种族：	二次电池
家族：	镍氢电池
适应温度：	-20～70 ℃

主要应用领域：太阳能照明、电动工具、电动玩具、家用电器、移动照明等。

Name:	Hydrogen Hero
Type:	Secondary Battery
Family:	Nickel-metal Hydride Battery
Operating temperature:	-20~70 ℃

Main application areas: solar lighting, electric power tool, electric toy, household appliances, mobile lighting, etc.

你不找他比一下吗?

Aren't you going to challenge him?

我更关注你，你们家族可是二次电池种族中的元老！

I am more interested in you. You know, your family are pioneers in the secondary battery field!

卧虎藏龙的竞技大赛，未知的强劲对手，这一切都使锂锂和大铅感到无比兴奋！于是，他们击掌相约——决赛再见！

Lithium Li and Big Lead were very excited. There were so many low-key strong opponents. They high-fived and agreed to meet in the final!

你应该知道，我在我们家族的地位有多高吧！

You know how high my status is in our family, don't you?

不吹牛，一些特定领域把最后的希望都寄托在了我身上！

No kidding! I'm the last hope for some fields.

我打包票，我工作的领域，你数都数不过来，手机、电脑、电动自行车、各种电动玩具……

I bet you can't even name those fields: phones, computers, e-bikes, and all kinds of electric toys…

我工作的地方，你仰望不及，电信、太阳能系统都离不开我。

My fields are beyond your reach. Both telecom and solar systems can't work without me.

有意思，那我们决赛再见！

Interesting! Then, see you in the final.

愿意奉陪！

Fine! Let's wait and see.

2

初赛——电池三项大比拼

Preliminary Round: The Big Battery Triathlon

转眼就到了初赛时间,裁判王一氯正在主席台上宣讲初赛内容和规则,通过了预选赛的选手们则在台下叽叽喳喳地讨论着。

It was the time for the preliminary round! Judge King Monochlorine was on the podium, explaining the competition's content and rules, while the contestants who had past the preliminary round were chattering with excitement below the stage.

姓名:　　　王一氯　　　
种族:　　　二次电池　　　
家族:　　　钠－氯化镍电池　　　
适应温度: 250～350 ℃
主要应用领域:电网储能、通信基站、动力电源、新能源汽车等。

Name: _____King Dichlorine_____
Type: _____Secondary Battery_____
Family: _____Sodium-nickel Chloride Battery_____
Operating temperature: _____250~350 ℃_____
Main application areas: _____grid energy storage, communication base stations, power supplies, and new energy vehicles, etc._____

对电池来说,最基础的考验就是容量和功率,这也是初赛的考核标准……

For batteries, the most fundamental tests are capacity and power—key criteria for the preliminary round.

姓名：锰叨叨
种族：一次电池
家族：碱锰电池
适应温度：-20 ~ 60 ℃
主要应用领域：军事通信装备、照相机、电动工具、电动玩具、游戏机等。

Name: Mnag
Type: Primary Battery
Family: Alkaline-manganese Battery
Operating temperature: -20~60 ℃
Main application areas: military communication equipment, cameras, electric power tools, electric toys, and gaming consoles, etc.

Name: Tyrant Nickel
Type: Secondary Battery
Family: Nickel-cadmium Battery
Operating temperature: -30~50 ℃
Main application areas: electric power tools, electric toys, emergency lights, portable cameras, and cordless telephones, etc.

姓名：镍霸
种族：二次电池
家族：镍镉电池
适应温度：-30 ~ 50 ℃
主要应用领域：电动工具、电动玩具、应急灯、便携式照相机、无绳电话等。

这对我们来说小菜一碟！
It's a piece of cake!

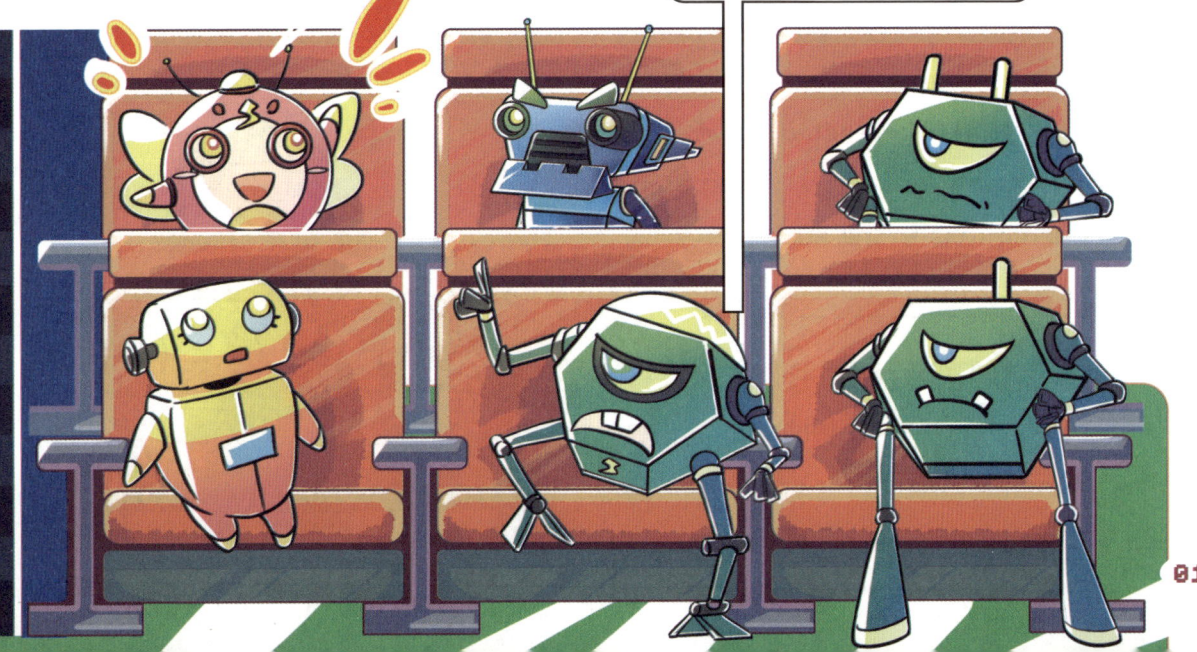

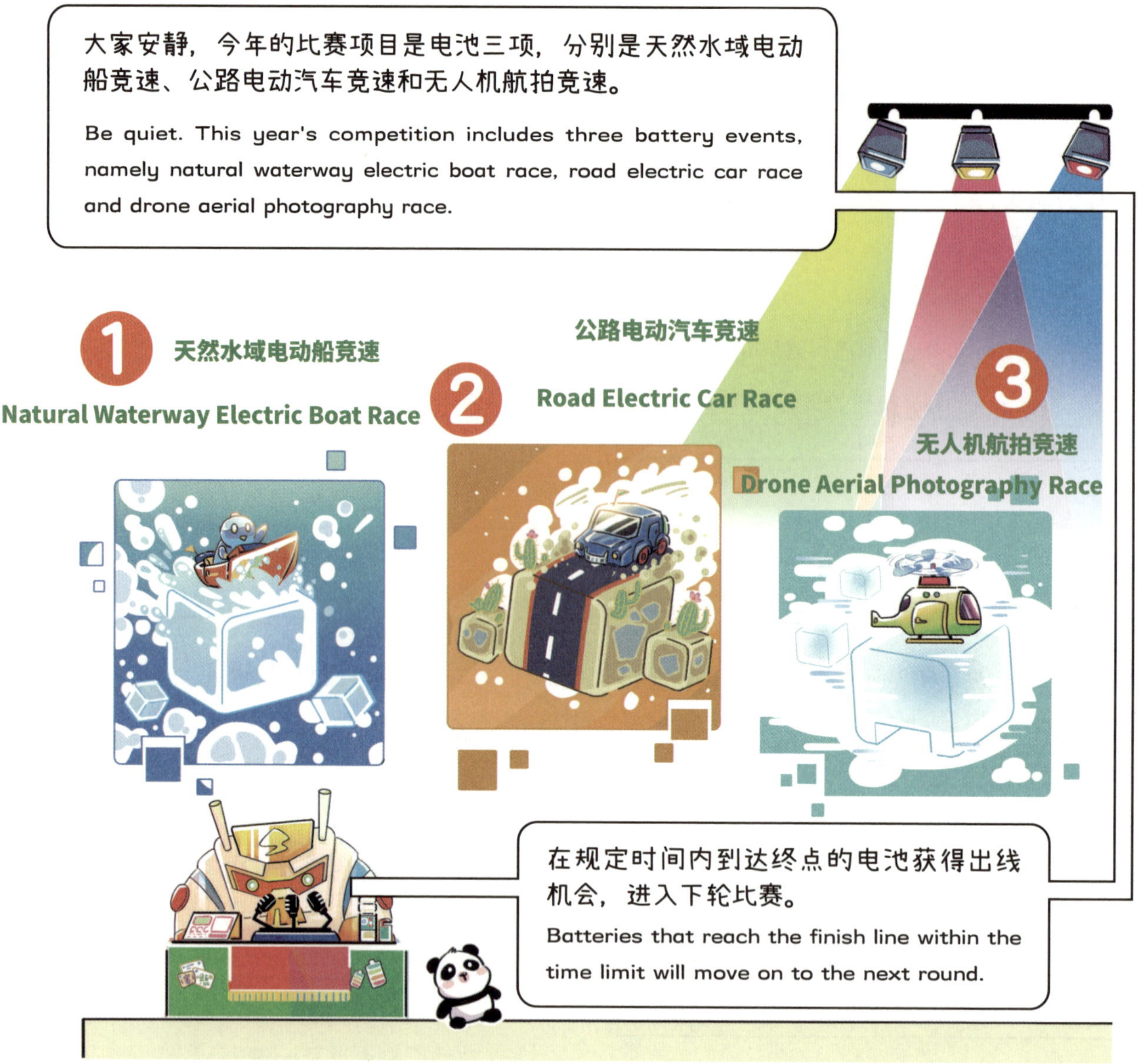

> 大家安静，今年的比赛项目是电池三项，分别是天然水域电动船竞速、公路电动汽车竞速和无人机航拍竞速。
>
> Be quiet. This year's competition includes three battery events, namely natural waterway electric boat race, road electric car race and drone aerial photography race.

1 天然水域电动船竞速
Natural Waterway Electric Boat Race

2 公路电动汽车竞速
Road Electric Car Race

3 无人机航拍竞速
Drone Aerial Photography Race

> 在规定时间内到达终点的电池获得出线机会，进入下轮比赛。
>
> Batteries that reach the finish line within the time limit will move on to the next round.

　　在裁判的示意下，大家逐渐安静下来。观众席上，只有镍霸还在和他的跟班们说着悄悄话……

　　As the judge signaled for silence, the crowd gradually quieted down. Only Tyrant Nickel was still whispering with his minions in the audience…

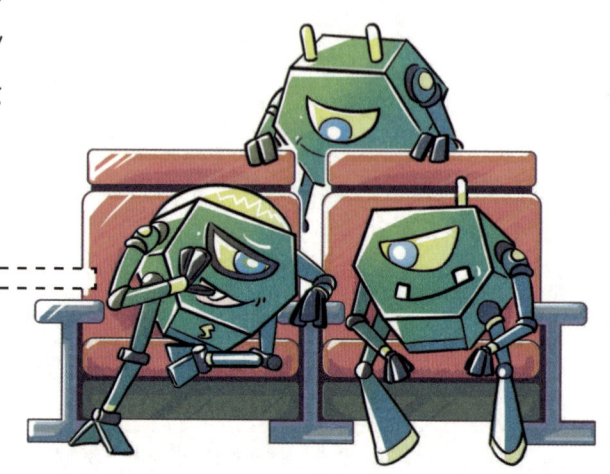

> 明白了，你们跟着我的手势行动。
>
> Got it? Follow my lead.

016

初赛第一场：天然水域电动船竞速。

赛场在一条宽阔的河流之上，从一座码头出发，到河流转弯处的另外一个码头结束。

Preliminary Round 1: Natural Waterway Electric Boat Race
The first event kicked off on a wide river. The course started at one dock, stretched along the river, and ended at another dock at the river bend.

初赛第一场：天然水域电动船竞速
Preliminary Round 1: Natural Waterway Electric Boat Race

裁判一声令下，初赛第一场比赛正式开始！

由锂锂驾驶的8号电动船如箭离弦，刚一开始就显现出了勇者无敌的势头。由大铅驾驶的9号电动船紧跟其后，其他电动船也不甘示弱地紧紧跟随。

赛事如火如荼，稳居第一、二名的8号电动船和9号电动船仍风驰电掣般向终点飞奔，此时已与其他船只拉开了一段距离，整个比赛演变成锂锂与大铅两位选手之间的对决。

At the referee's signal, the first round of the preliminaries was officially underway!

Lithium Li, piloting Boat No.8, launched like an arrow from the starting dock, displaying an unstoppable momentum right from the start. Hot on their heels was Big Lead, steering Boat No.9, followed closely by the rest of the determined competitors.

The race quickly heated up. As the boats charged forward, Boat No.8 and Boat No.9 maintained their lead, speeding toward the finish line and leaving other boats trailing behind. The competition intensified, evolving into a head-to-head duel between Lithium Li and Big Lead.

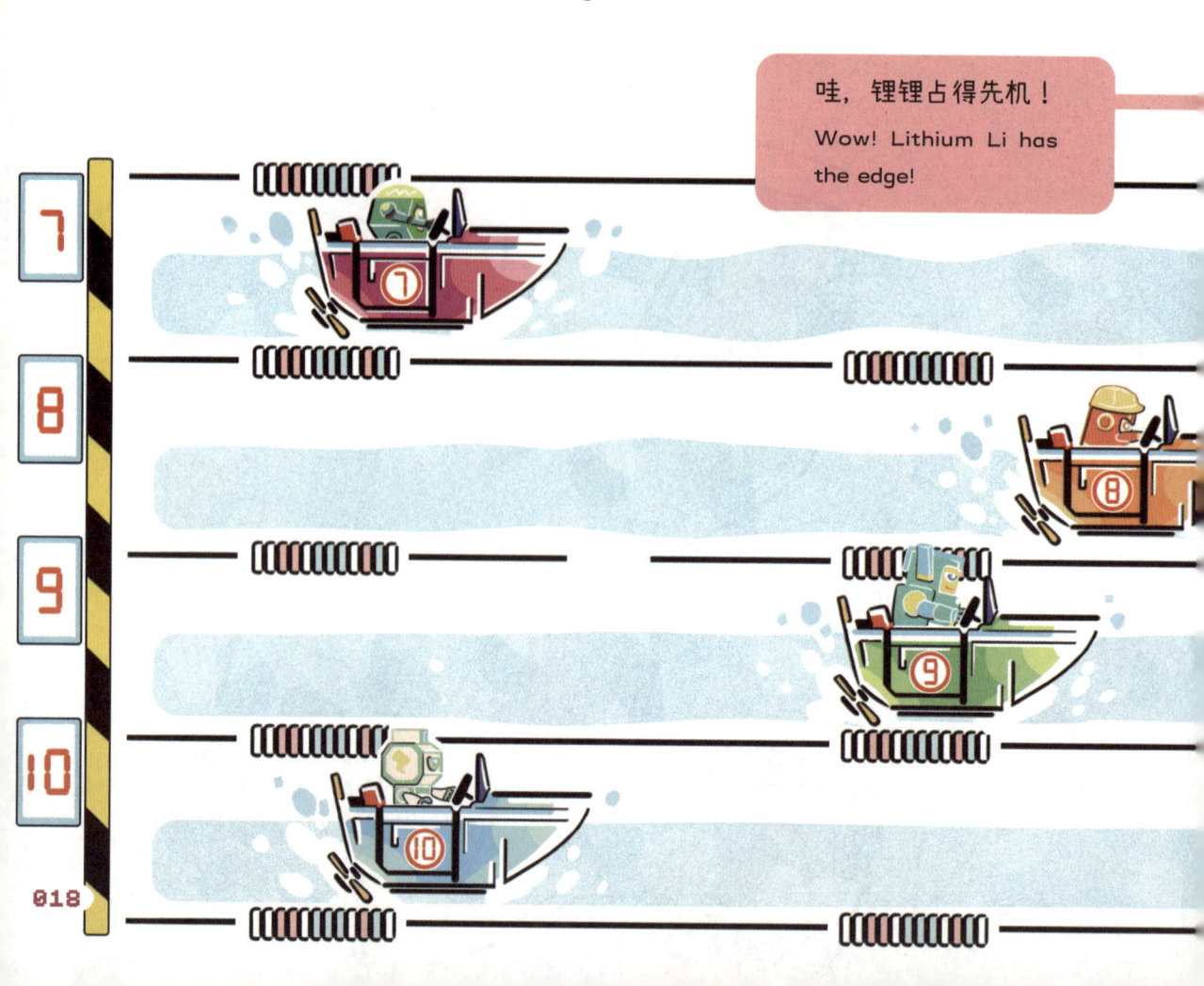

哇，锂锂占得先机！
Wow! Lithium Li has the edge!

虽然有些许落后，大铅并不气馁，仍专注驾驶，奋力追赶。到了河流中游，8号电动船与9号电动船的距离越缩越短，渐渐地，9号电动船超越了8号电动船。

Although falling slightly behind, Big Lead remained undeterred, focusing intently on steering and pushing forward with determination. By the time they reached the middle stretch of the river, the gap between Boat No.8 and Boat No.9 narrowed. Inch by inch, Boat No.9 pulled ahead, overtaking Boat No.8.

呀，大铅后来者居上了！
Wow! Big Lead has taken the lead!

即将冲刺终点了，赛事进入白热化阶段！锂锂在最后一个弯道瞅准时机突然发力，只是瞬间，8号电动船和9号电动船又并驾齐驱了。

As the boats went near the finish line, the race reached a fever pitch. Entering the final turn, Lithium Li seized the moment, suddenly surging forward with a burst of speed. In an instant, Boat No.8 and Boat No.9 were advancing in line, locked in an electrifying duel.

随着裁判的令旗挥动，一场你争我夺的比赛结束了。最终，8号电动船第一个靠岸。

With the referee's flag waved, the intense battle came to an end. Boat No.8 was the first to reach the dock, claiming victory in this fierce race.

锂锂第一名，大铅第二名，锌博士第三名……

1st Place: Lithium Li
2nd Place: Big Lead
3rd Place: Dr. Zinc
…

初赛第一场，锂锂以领先0.01秒的成绩险胜大铅，拿下第一名。

In Preliminary Round 1, lithium Li narrowly defeated Big Lead by a margin of 0.01 seconds, securing the first place.

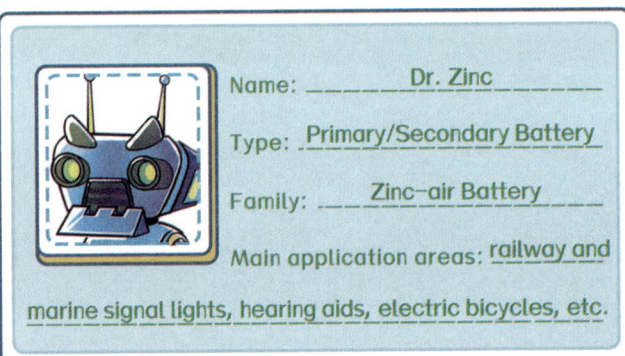

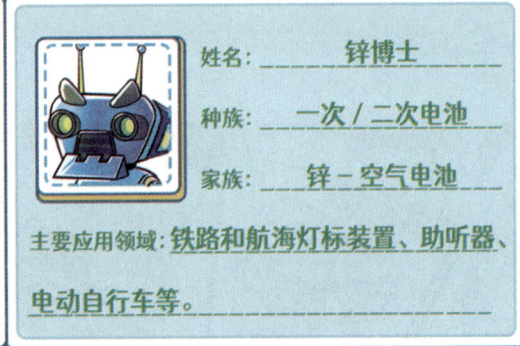

Name: Dr. Zinc
Type: Primary/Secondary Battery
Family: Zinc-air Battery
Main application areas: railway and marine signal lights, hearing aids, electric bicycles, etc.

姓名：锌博士
种族：一次／二次电池
家族：锌－空气电池
主要应用领域：铁路和航海灯标装置、助听器、电动自行车等。

后面的比赛我要取得更好的成绩！

I will do better in the upcoming races!

不要神气得太早，初赛还有2场，最后能赢才算数！

Don't get too cocky; there are two more preliminary rounds. Winning in the end is what counts!

初赛第二场：公路电动汽车竞速。

赛道是一条在山林之间蜿蜒盘旋的公路。

Preliminary Round 2: Road Electric Car Race

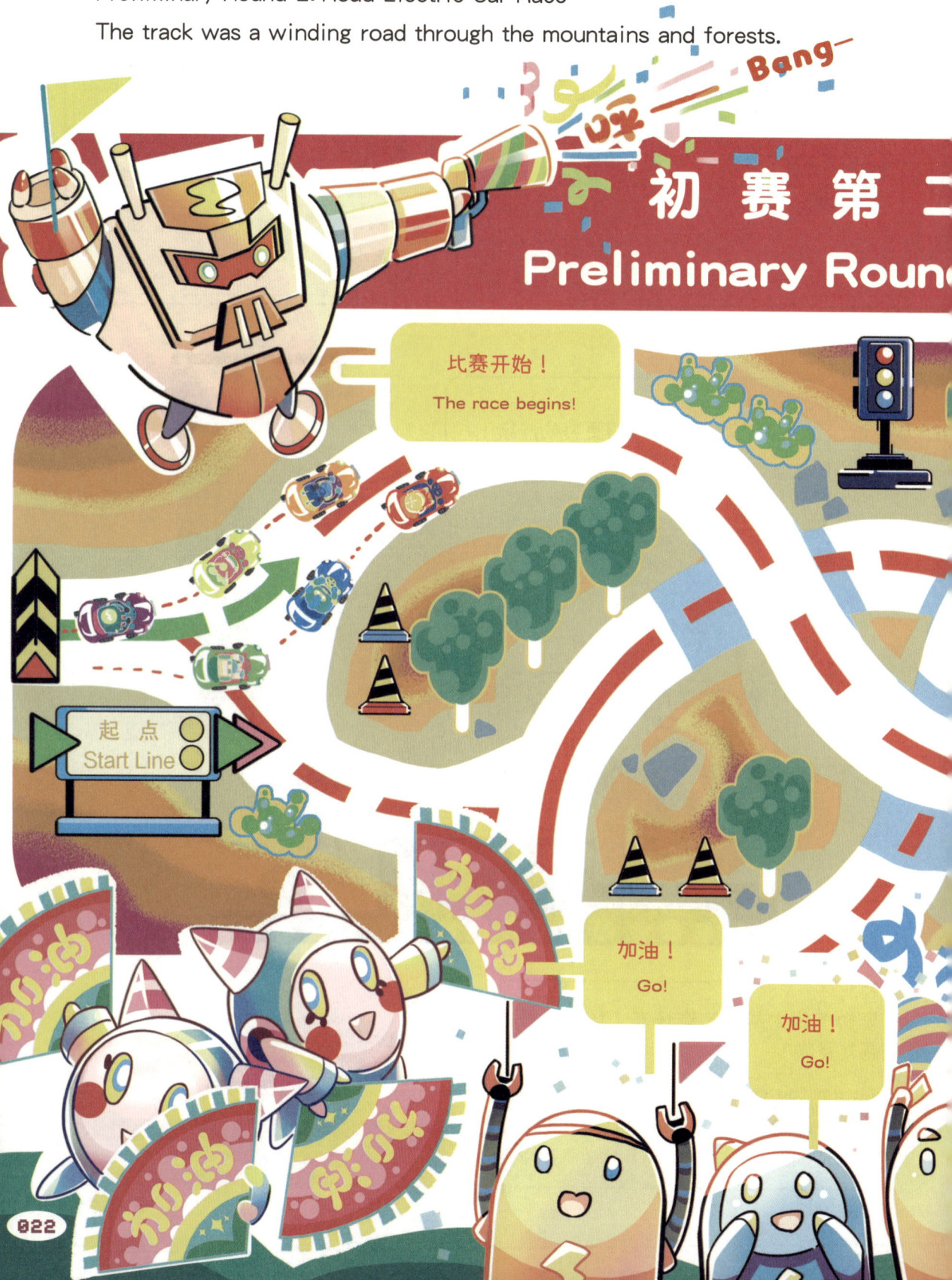

The track was a winding road through the mountains and forests.

比赛开始！
The race begins!

起点
Start Line

加油！
Go!

加油！
Go!

嗡嗡——比赛开始，场上所有电动汽车齐刷刷地冲出。啦啦队跳起了舞，观众席上的观众都站了起来，为选手们呐喊助威。

Vroom—The race began, and all the electric cars rushed out in unison. The cheerleaders started dancing, and the audience stood up, cheering and shouting to encourage the contestants.

场：公路电动汽车竞速
2 Match: Road Electric Car Race

终点
Finish Line

加油！加油！
Go! Go!

赛程刚开始，选手们刚进入第一个弯道，就发生了事故。本该在第三车道行驶的由镍霸驾驶的电动汽车，不知怎的竟脱离了自己的赛道，向一旁的第四车道直冲而去。

As the race began, just after the contestants the first turn, an accident occurred. The car driven by Tyrant Nickel, which should have been in the third lane, suddenly swerved out of its lane and veered into the fourth lane.

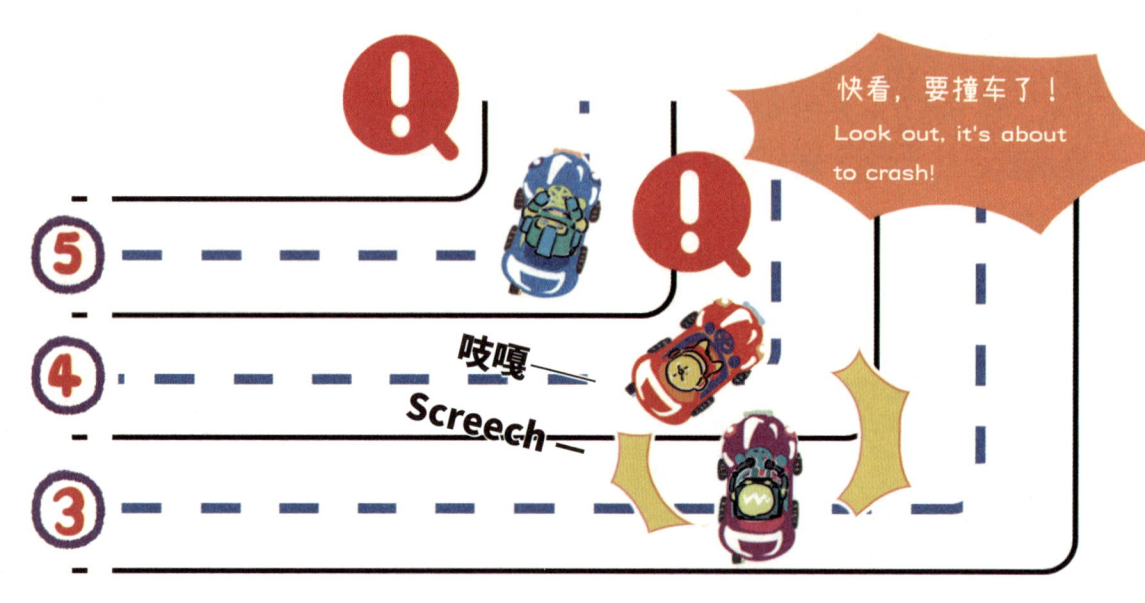

快看，要撞车了！
Look out, it's about to crash!

吱嘎
Screech

由锂锂驾驶的电动汽车躲闪不及，失去了控制，一下子撞到了正行驶在第二车道的大铅所驾驶的车上。

幸好大铅的安全性极好，这才没有电池人伤亡，只是锂锂的电动汽车车轮陷进了一个小坑里，发动机发出呜呜的声音，无法继续行驶。

大铅见状，并没有立即重新启程，而是选择了帮助锂锂。他们合力将车从坑里弄了出来，这才继续上路。

The electric car driven by Lithium Li failed to dodge and lost control, crashing into the car driven by Big Lead who was running in the second lane.

Luckily, Big Lead's car had excellent safety features, so no one was hurt. However, Lithium Li's car got stuck in a small pit, and the engine started making a whining noise. Lithium Li was unable to continue the race.

Seeing this, Big Lead didn't immediately start again. He chose to help Lithium Li. They managed to get the car out of the pit together before continuing the race.

而身为罪魁祸首的镍霸仿佛只是晃了一下神，在酿成这场事故之后瞬间回归了自己的赛道，火力全开继续向终点奔去。

　　选手们陆续抵达终点。果然，锂锂和大铅因为这次事故落后许多，锌博士、氢天天、镍霸夺得了前排名次。

Meanwhile, the culprit, Tyrant Nickel, seemed barely distracted by the accident. After causing the crash, he quickly returned to his lane and raced ahead with full throttle toward the finish line.

One by one, the contestants crossed the finish line. As expected, Lithium Li and Big Lead fell far behind due to the accident, while Dr. Zinc, Hydrogen Hero, and Tyrant Nickel took the top three spots.

> 锌博士、氢天天、镍霸分别位列前三，第四名是……
> Dr. Zinc, Hydrogen Hero, and Tyrant Nickel claimed the first, second, and third place, respectively. The fourth place was…

听说发生事故了？
It's said that there was an accident.

谁会在意呢？得分排名已经出炉了！瞧，我多受欢迎！
Who cares? The scores and rankings are out now! Look how popular I am!

比赛结束后，锂锂为表感谢，邀请大铅到自己家中一起用晚餐。饭桌上，两人聊到今天的比赛，都不免显得有些垂头丧气。

机器人X来到桌前，播放了他对这场事故原因的调查结果，大家这才恍然大悟，原来都是镍镉电池家族在捣鬼！

After the race, Lithium Li invited Big Lead to his home for dinner to show his gratitude. At the dinner table, the two talked about the race, both feeling a bit downhearted.

Robot X came over to the table and played the results of his investigation into the cause of the accident. Everyone suddenly realized that the Nickel-cadmium Battery family played tricks!

事故原委就是这样了……
Here's how the accident happened…

在接下来的比赛中，我们都要更加小心！
In the upcoming races, we all need to be more cautious.

原来如此！这一切都不是偶然！
So that's what happened! It wasn't a coincidence!

镍镉电池家族太可恶了！
The Nickel-cadmium Battery family is so despicable!

怎么能这样？
How could they do such a thing?

下一场比赛是无人机航拍竞速，为了以防万一，锂锂和大铅决定为他们的无人机临时加装紧急避障系统，这样或许能够应对镍镉电池家族的攻击。

The next race was drone aerial photography race. To be cautious, Lithium Li and Big Lead decided to install emergency obstacle avoidance systems on their drones, just in case that the Nickel-cadmium Battery family might sabotage them again.

> 这样应该能有帮助！
> This should help!

> 真希望他们能够遵守规则！
> I really hope they'll follow the rules!

第三场比赛：无人机航拍竞速。

设置于户外一片开阔的草地上，各式各样的自制无人机使人目不暇接。

The Third Round: Drone Aerial Photography Race

The race was in an open field outdoors, where various homemade drones were dazzling.

所有无人机从起降台起飞后，上升到 10 米高度，水平飞向目标物，离目标物 2 米距离开始拍摄。

All the drones take off from the launch pad, climb to a height of 10 meters, and fly horizontally toward the target. Once within 2 meters of the target, they should begin capturing images.

拍摄内容包括目标物的四个侧面和俯视角度，然后返回起降台，结束计时。

The task includes taking photos from four sides and a top-down view of the target. Once done, drones should return to the launch pad to stop the timer.

起降慢，返回迟，降落在起降台外，或者拍摄图像不清晰的都要扣分。总分 10 分。

Any issues like slow take-off, delayed returns, landing outside the pad, or unclear images will result in point deductions. The maximum score is 10 points.

10m

起降台
Launch Pad

初赛第三场：

Preliminary Round 3:

这场比赛对选手与设备的要求都相当高，飞行当中除了需要满足安全性与稳定性的条件以外，同时还要兼顾速度、反应度和灵活度，才能取得优异的成绩。

The race was a tough one for the contestants and their equipment. Besides ensuring safety and stability during the flight, one must also consider speed, reaction, and agility to achieve optimal performance.

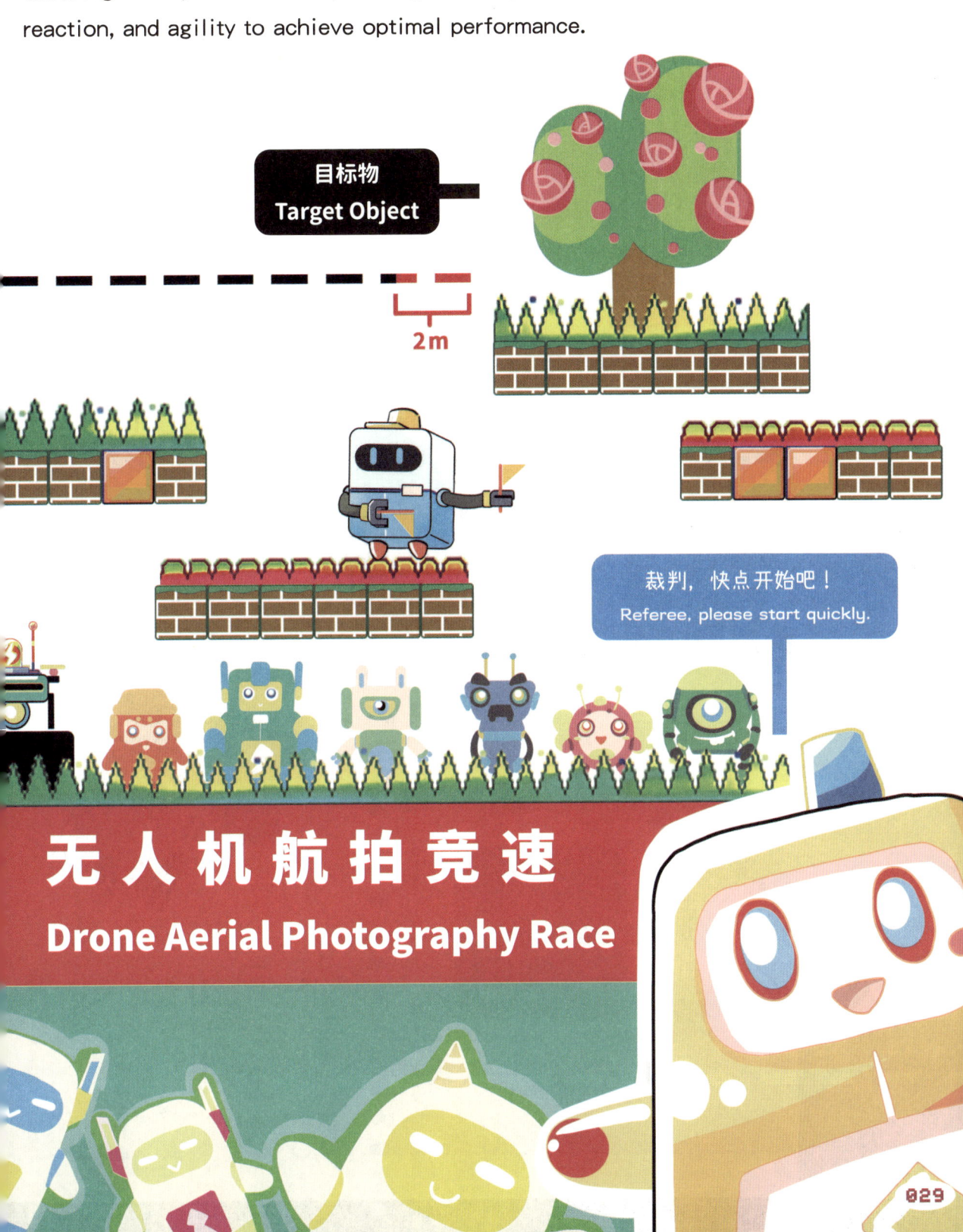

目标物
Target Object

2m

裁判，快点开始吧！
Referee, please start quickly.

无人机航拍竞速
Drone Aerial Photography Race

随着发令信号响起，起降台上所有无人机应声起飞，嗡嗡的振动声此起彼伏。

不同于前两场有开阔赛道的比赛，这更像是一场混战！有一些表现优异的无人机能够灵敏地冲出重重机群的包围，到达目标物附近；另一些就没那么幸运了，在飞行途中就跌落到草地上，直接丢掉本场比赛资格。

With the starting signal, all drones on the launch pad took off, and their buzzing vibrations filled the air.

Unlike the previous two races with open tracks, this one resembled a chaotic battle! Some exceptional drones were able to nimbly break through the crowd of contestants and reach the target area. However, the less fortunate ones fell to the ground during the flight, instantly being disqualified from the race.

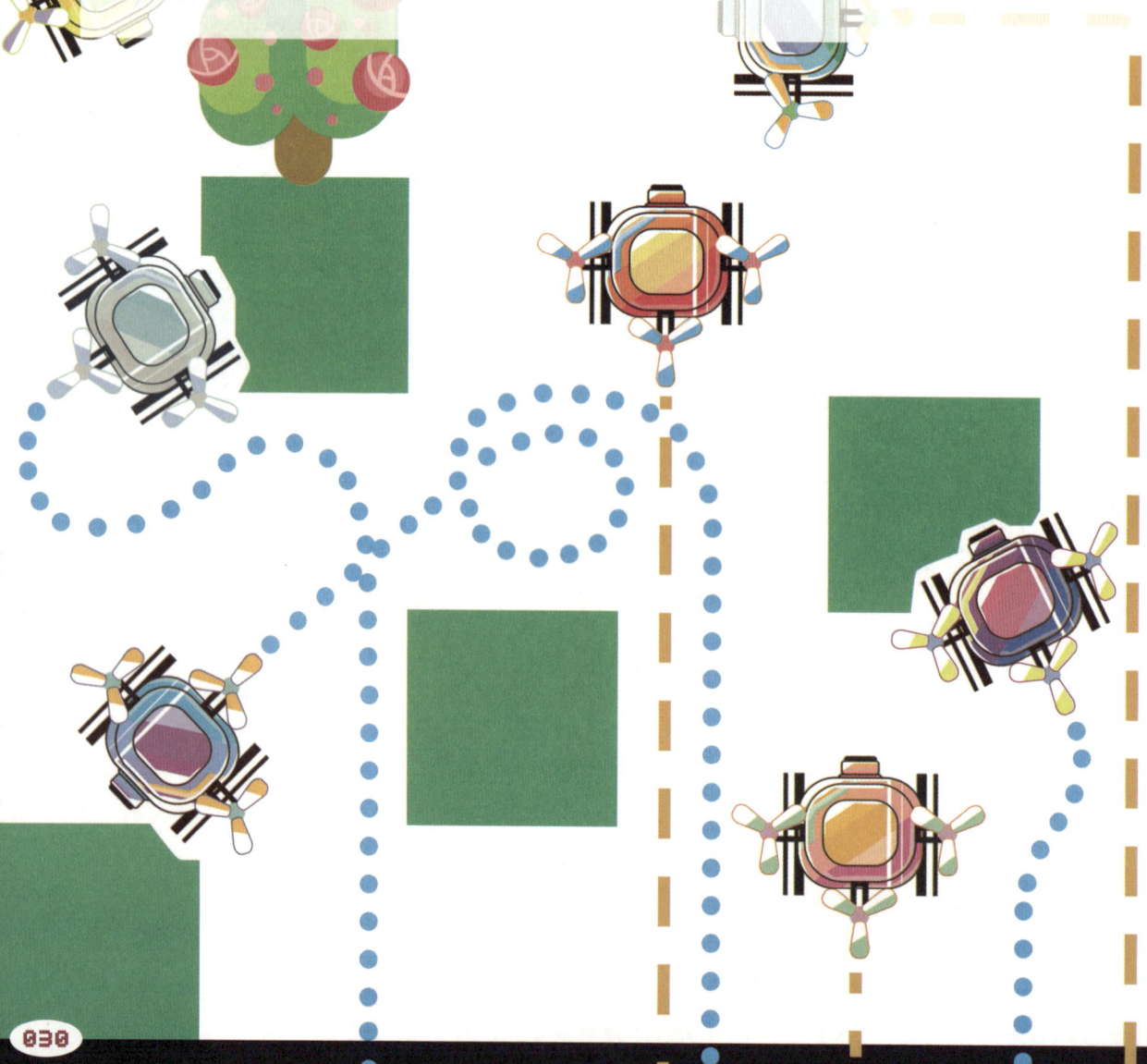

哪个角度更好呢？

Which angle is better?

　　锂锂遥控着无人机也顺利到达了目标物附近，锂锂正在专心拍摄，没有注意到身后有架无人机正在悄悄靠近……

　　突然，由锂锂所遥控的无人机猛然晃动了一下，是昨晚刚刚安装的紧急避障系统生效了！镍霸遥控的无人机猛然撞击而来，触发了锂锂所驾驶的无人机的感应功能，锂锂的无人机自动闪避到安全区域去了。

Lithium Li controlled the drone and reached the target area. Lithium Li was so focused on capturing images that he didn't notice a drone approaching from behind quietly…

Suddenly, Lithium Li's drone jerked to one side. The system installed the night before took effect! The drone controlled by Tyrant Nickel collided with Lithium Li's drone, triggering the sensor on Lithium Li's drone. Immediately, Lithium Li's drone automatically dodged into a safe zone.

附近正在拍摄的电池人们都留意到了镍霸这种没有竞技精神的行为，不约而同地抵制起来，只为守护心中的正义，维持比赛的公开、公正、公平！

The battery participants filming nearby all noticed the lack of sportsmanship of Tyrant Nickel. They all united to protect the justice, and maintain the openness, fairness, and impartiality of the competition.

快向我求助吧！
Come to me for help!

犯规者我不喜欢，拍完左边，拍右边；拍完右边、拍后边；还有上面……
I don't like cheaters. Left side first, then right side; after right side, the back; and then the top…

速战速决，我已经拍完了。
Let's finish it quickly. I'm done with my shots.

要仔细，要认真才能拍出高清照片……拍完，我来帮你！
Be careful. Only then will you capture high-quality photos… I'll help you once you're done!

团结一致，制止犯规者！
United as one, we'll stop the rule-breaker!

摆脱了镍霸的骚扰，锂锂遥控着无人机顺利地飞向目标物上方并成功拍好了照片。

Free from the harassment of Tyrant Nickel, Lithium Li remotely controlled the drone and flew toward the upper side of target smoothly, capturing the photo with precision.

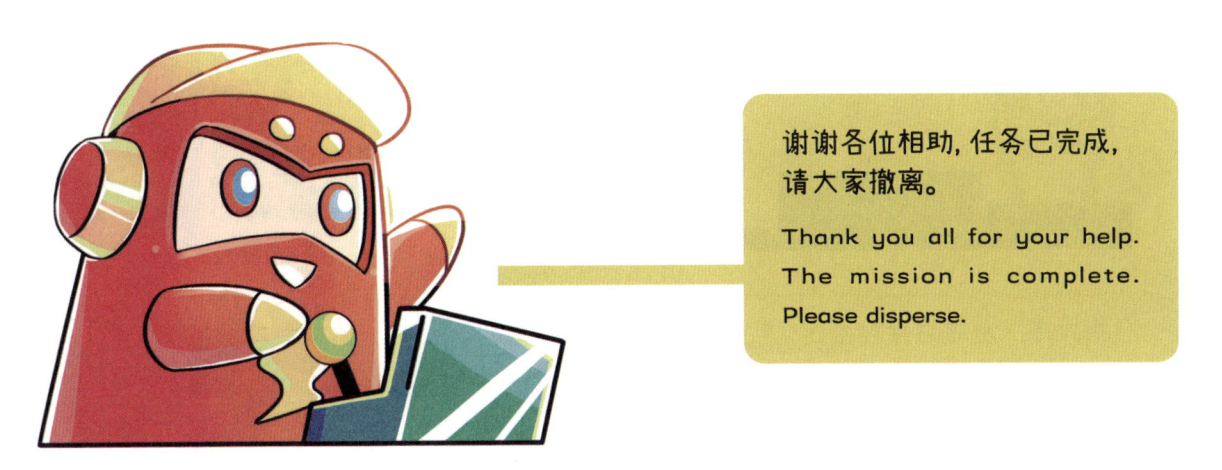

谢谢各位相助，任务已完成，请大家撤离。

Thank you all for your help. The mission is complete. Please disperse.

计时结束，选手们遥控无人机陆陆续续降落到起降台上。

As the timer stopped, the contestants gradually landed their drones back on the launch pad.

你们都是好样的！为你们点赞！

You all did great! Thumbs up for you!

初赛的最后一场比赛结束了，不少选手的无人机在限时内还没进入起降台范围，就不得不滑行降落了。

The final race of the preliminary round ended, and many contestants' drones failed to land on the designated area of the platform within the time limit, forcing them to make a glide landing instead.

锰叨叨第一，镍霸第二，锂锂第三……
Mnag finished first, Tyrant Nickel second, and Lithium Li third...

飞得太高，我难以保持稳定，功率不大，真伤脑筋！
I flew too high; it was hard to maintain stability. My power was insufficient—it's really nerve-wracking!

最后10米，我的电量耗尽了！
In the last 10 meters, my battery ran out!

那个镍霸，绝对是故意的！
That Tyrant Nickel definitely did it on purpose!

我们观众的眼睛可是雪亮的！
We, the audience, can see it!

今天真是太感动了！
What happened today was truly moving!

我们电池王国都像这样团结起来，才能有更好的未来！
We Battery Kingdom can only have a better future if we stay united like this!

请各位选手暂时不要离场，到大赛休息室等候裁判宣布晋级赛比赛安排。
All contestants please remain in place and wait in the competition lounge for the judge to announce the arrangements for the next round of the competition.

在一片争议与感动当中，众电池人陆陆续续走出了赛场，来到休息室，纷纷对镍镉电池谴责起来。

Amidst the controversy and emotion, the battery participants gradually walked out of the arena and into the lounge, where they began to condemn the Nickel-cadmium Battery family.

各位选手，公开、公正、公平是电池竞技大赛的基本原则。

Dear contestants, fairness, justice, and impartiality are the fundamental principles of the Battery Competition.

那种靠影响比赛判罚而取得的胜利，有损于竞技精神，注定不会得到观众由衷的欢呼和掌声。

Victory gained by influencing the competition undermines the spirit of the competition and will never earn the genuine cheers and applause of the audience.

操控无人机都有失手的时候，凭什么认为我攻击他嘛！

Everyone makes mistakes when controlling drones. Why did you think I attacked him?

不以为耻，反以为荣！镍霸，你还好意思说！

Far from being ashamed of it, you glories in it! Tyrant Nickel, how dare you say it!

今天还好有大家的帮助，想赢比赛，要靠实力，不能破坏规则。

It was really lucky to have your help. To win the game, one must rely on one's own abilities and not break the rules.

有输有赢、充满悬念正是比赛的精彩之处。我宣布，初赛到此结束！

Winning or losing, the competition is full of suspense! That's what makes a competition exciting. I announce that the preliminary rounds are now over!

晋级赛入围者需要根据三场比赛的综合成绩判定，我们将在计算完每位选手的得分后公布晋级选手名单。请大家好好休息，留意比赛通知！

Contestants moving to the next phase will be determined based on the overall performances from the precious three races. After we tally each contestant's scores, we will announce the list of those who will move on. Please take a good rest and stay tuned for the competition updates!

好的，裁判！

Understood, judge!

3

极具挑战的晋级赛

Challenging Advancement Race

清晨，睡梦中的锂锂被一阵轻音乐吵醒，机器人 X 突然放声歌唱起来。原来是初赛结果公布了，锂锂成功晋级！整个房间都荡漾着欢呼声与祝贺声。

In the early morning, Lithium Li woke up from soft music as Robot X suddenly burst into singing. The results of the preliminary race had been announced: Lithium Li had made it to the next round! The room was buzzing with cheers and congratulations.

> 嗨，家人们，有人过生日吗？
> Hey, guys! Is it anyone's birthday?

> 是个好消息，恭喜你入围晋级赛了。
> Good news. Congratulations on your promotion.

> 太棒了！太棒了！
> That's awesome!

> 恭喜！
> Congratulations!

> 在即将到来的晋级赛中，我们还新增了安全性、灵活性、适应性等作为考验对象，这也正是电池王国各个家族正面临巨大挑战！
> Safety, flexibility, and adaptability will be the newly added marking criteria in the advancement race. These are big challenges that all the families in the Battery Kingdom are facing!

锂锂瞬间清醒，一个跟头就从床上翻坐起来，来到电视机跟前仔细观看，得知晋级赛的驾驶装置竟然是由选手现场自由组装！

Lithium Li was immediately wide awake, flipping out of bed and rushing to the TV to watch closely. It showed that the contestants had to assemble their driving devices on the spot for the advancement race!

晋级赛——障碍越野大赛
Advancement Race—Obstacle Cross-country Race

每位选手驾驶由各自现场组装的装置依次穿越石林、沙漠、冰川。三种环境中分别设置有不同的障碍，前四名到达终点的选手将获得进入决赛的资格。

Every contestant drives his/her vehicle assembled on-site. They need to travel through the stone forest, desert, and glacier, each with different obstacles. The top four finishers will be qualified for the finals.

等等，自由组装？晋级赛有多少人参加呢？

Wait, on-site assembly? How many contestants are there in this round?

是的。共有64名选手入围晋级赛。

Well. Sixty-four contestants were selected for the advancement race.

晋级赛前夜，各路选手仍在为组装驾驶装置费尽脑力。

The night before the race, everyone was trying hard to prepare to put their driving devices together.

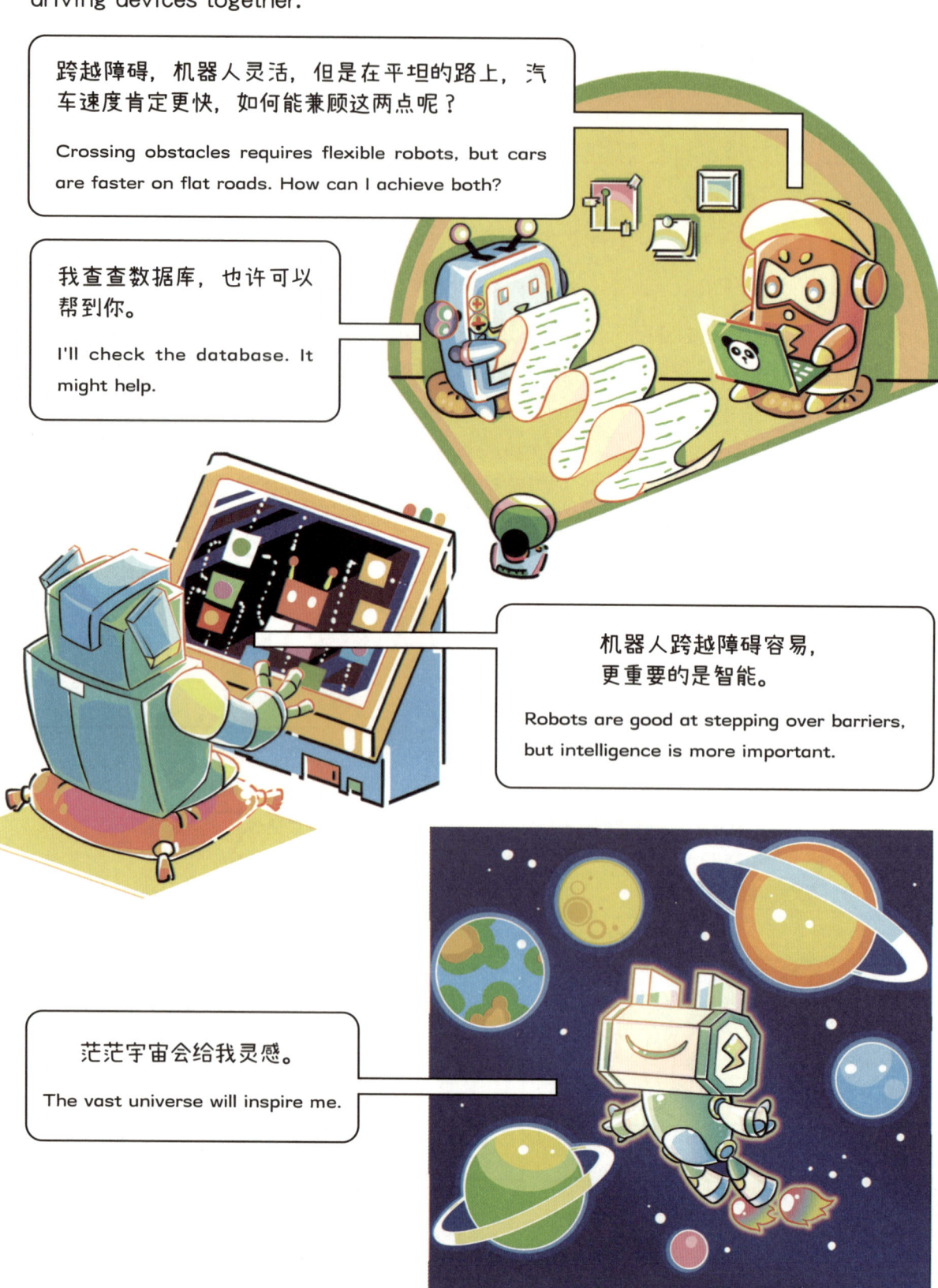

跨越障碍，机器人灵活，但是在平坦的路上，汽车速度肯定更快，如何能兼顾这两点呢？

Crossing obstacles requires flexible robots, but cars are faster on flat roads. How can I achieve both?

我查查数据库，也许可以帮到你。

I'll check the database. It might help.

机器人跨越障碍容易，更重要的是智能。

Robots are good at stepping over barriers, but intelligence is more important.

茫茫宇宙会给我灵感。

The vast universe will inspire me.

我相信书本的力量，组装对我锌博士来说一点儿也不复杂！
I believe in the strength of books! For me, Dr. Zinc, assembly is a piece of cake!
电池原理

这个装置要怎么弄，我真的不擅长啊！
How can I put these things together? I'm not good at this!

快给我找最好的设计方案，我一定要组装出最好的座驾。
Quickly, find me the best design! I will build the best ride!

转眼就来到第二天，随着离赛事开始的时间越来越近，选手们愈发斗志激昂，在万众瞩目下走进了赛场。

The next day arrived. As the start time approached, the contestants got more motivated. They entered the arena under the attention of the masses.

电池竞技大赛
Battery Competition

电池竞技大赛・晋级赛——障碍越野大赛赛道图
Battery Competition: Advancement Race—Track Map of Obstacle Cross-country Race

都不给选手们歇口气的时间！这不符合常规！
They don't even give contestants a break! That doesn't make sense!

作为五年一度的国家级赛事，如此节奏紧凑的赛程与逐渐拔高的难度引起了部分观众的质疑。

As a national-level event held every five years, the tight schedule and increasing difficulty have raised doubts among some audience.

晋级赛
Advancement Race

障碍环境——石林、沙漠、冰川
Obstacle Environments—Stone Forest, Desert, Glacier

一定是发生了什么事情！
There must be something happening.

我听说，电池王国要打仗了！
I heard there is going to be a war in Battery Kingdom!

进入赛场前，选手们首先来到的是赛前准备室，他们需要在这里组装好自己的驾驶装置。这一步骤大家都早有准备，显得胸有成竹，但锰叨叨除外。

Before entering the arena, the first stop for contestants was the preparation room. Contestants should assemble their driving devices here. Everyone seemed confident, except for Mnag.

这次晋级赛将是一场难以预测的超级障碍越野赛，更是一场对极限的挑战！

It's going to be an unpredictable super obstacle cross-country race—a race that will challenge the limit!

好啦！大功告成！

All done! Perfect!

为了激励更多优秀选手完赛，我们还特设了"金头盔奖"，这个奖励会颁发给所有在规定时间内到达终点的选手。

To encounter more outstanding contestants to finish, we've also introduced the "Golden Helmet Award" to award all who reach the finish line in the allotted time.

下面请利用电脑，在30分钟内组装好自己的驾驶装置，超时未完成的选手直接淘汰。

Now, please use the computer to assemble your driving devices in thirty minutes. Contestants who fail to complete it in time will be directly eliminated.

30分钟很快过去，大多数选手都已组装完毕，只有极个别选手没有完成，直接被淘汰了。

　　众选手来到了工作仓区域，大家都选择的是机器人，唯独锂锂径直走向了一辆电动汽车，这引起旁边的镍霸一顿嘲讽。

Thirty minutes flew by. Most contestants had completed, with only a few being eliminated for not finishing in time.

They moved to the working area, and they all chose robot. However, Lithium Li went straight to an electric car, which drew mockery from Tyrant Nickel nearby.

> 究竟谁会掉坑里，让我们拭目以待吧！
> Let's see who ends up stuck!

> 哈哈，竟然选择汽车，小心再掉坑里，出不来！
> Nice choice, huh? Be careful not to get stuck in a pit again!

啪！随着信号枪发出响声，越野正式开始。
Bang! With the sound of the starting gun, the race began.

哇！锂锂真可谓是一骑绝尘呐！
Wow! Lithium Li is leaving everyone in the dust!

电动汽车在平路上快，待会儿越野就傻眼了！
Electric cars run fast on flat roads, but they'll be in trouble in the cross-country section!

由于起点距离石林路段还有一段距离，这一赛程是平坦的大道，选择驾驶电动汽车的锂锂显现出巨大优势，一路狂飙，遥遥领先于其他选择驾驶机器人的选手。有的选手不想落后太多，宁愿消耗大量电能，也要使用"超级加速"功能努力追赶。

Since there was a stretch of smooth road between the starting point and the stone forest section, driving an electric car gave Lithium Li a significant advantage. He speeded ahead of other contestants who chose robots. Some contestants were unwilling to fall too far behind so they activated the "super acceleration" function to catch up, even if that could consume considerable energy.

快点！再快点！
Faster! Faster!

锂锂驾驶着电动汽车第一个到达了石林路段，望着眼前泥泞的沙石、鹅卵石纵横的路面，他果断按下了改变驾驶形态的按钮！

Lithium Li was the first to reach the stone forest section. Seeing the muddy and rocky road ahead, he decisively pressed the button to change the driving mode!

难以置信，竟然会变形！

Incredible! It can transform.

是我看错了吗？锂锂的车正在飞速的重组！

Am I wrong? Lithium Li's car is rapidly transforming!

快点！再快点！不能浪费优势！

Faster! Faster still! Don't waste the lead!

咔——咔——

电动汽车进入重组状态，有几个机器人渐渐赶了上来。

Click—Click—

As the electric car was reconfiguring, several robots gradually caught up.

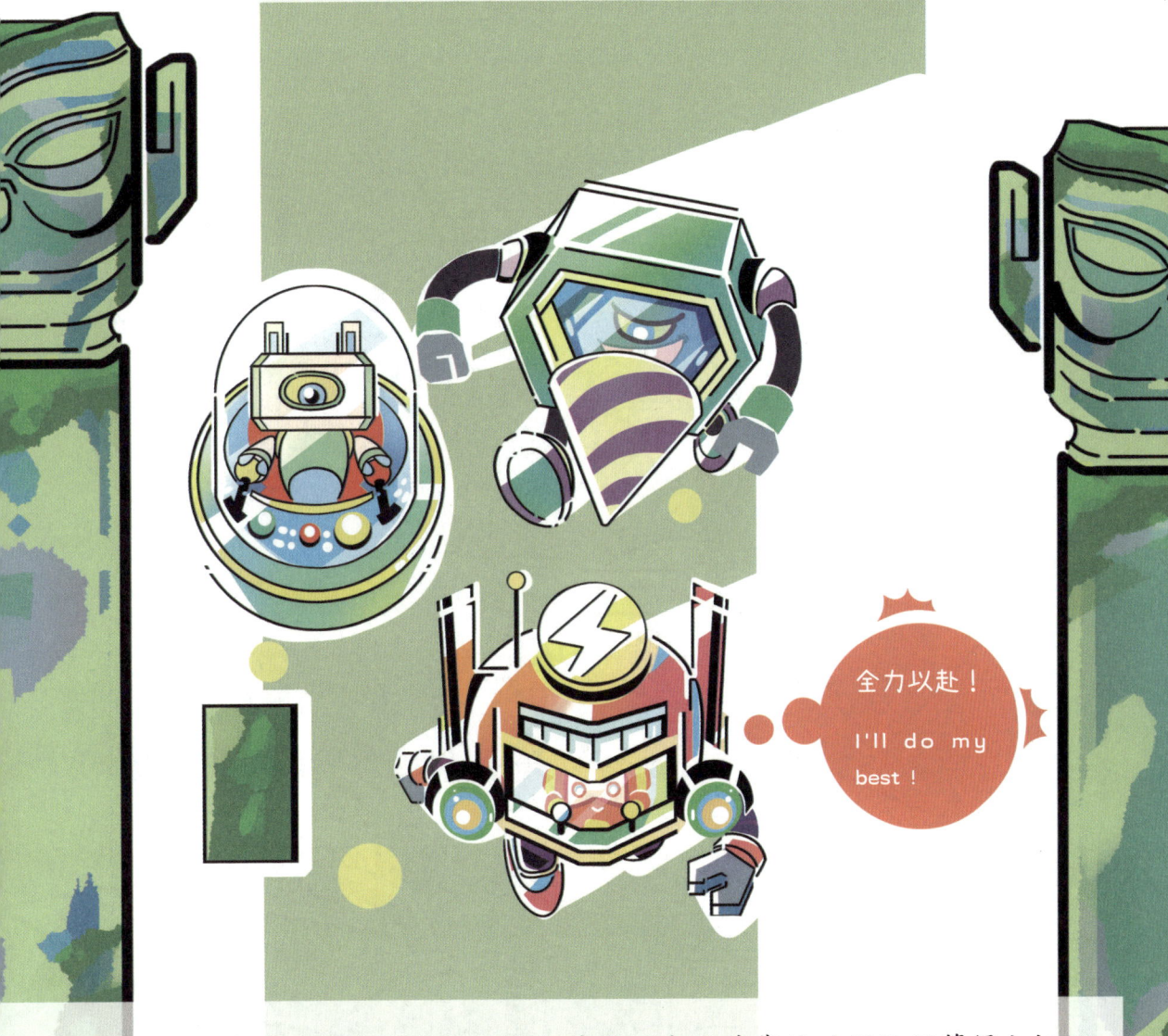

全力以赴！
I'll do my best！

终于重组完毕。虽然此次锂锂改变装置形态所花费的时间已经算得上极短了，但机器人们"超级加速"的势头也很猛！此时进入石林，其他选手与锂锂的差距已不大。

Finally, it was done. Although the time it took Lithium Li to transform was almost the shortest, other contestants speeded up quickly with "super acceleration"! As Lithium Li entered the stone forest, the gap between the others and him was narrowed.

石林的挑战有目共睹，不仅有大自然为选手们准备的天然"刑具"——例如由山脉上自上而下的流水形成的一道道河沟和泥泞的地面，以及无数光滑的鹅卵石，还有赛事举办方搭建的木桩、铺散的碎石块等障碍物。

The challenges the stone forest presented were evident. There were natural "traps" like streams formed by mountain runoff, muddy road, and countless smooth pebbles, along with obstacles like stakes and rocks set up by the organizers.

哎哟！
Ouch!

嘿嘿！走自己的路，让别人无路可走！
Hehe! Go my own way, and leave others with none!

很多机器人倒在了这条障碍重重的路上，它们关节断裂、传感器报废、电力驱动器损坏……有近1/3的机器人在此折戟。

Many robots fell on this tough course, with broken joints, damaged sensors, malfunctioning power drives, etc. Nearly one-third of them were defeated here.

前方已经隐隐约约能看到沙漠的影子了，锂锂所驾驶的机器人仍然保持第一。在充满未知障碍的赛段上打头阵是相当危险的一件事，于是，有一些机器人不再另外开辟道路，而是紧跟着锂锂的步伐。

The desert was faintly visible ahead and Lithium Li's robot remained in the lead. The section was filled with unknown obstacles. Leading the way here was quite dangerous. Instead of forging new paths, some contestants thus followed in Lithium Li's footsteps.

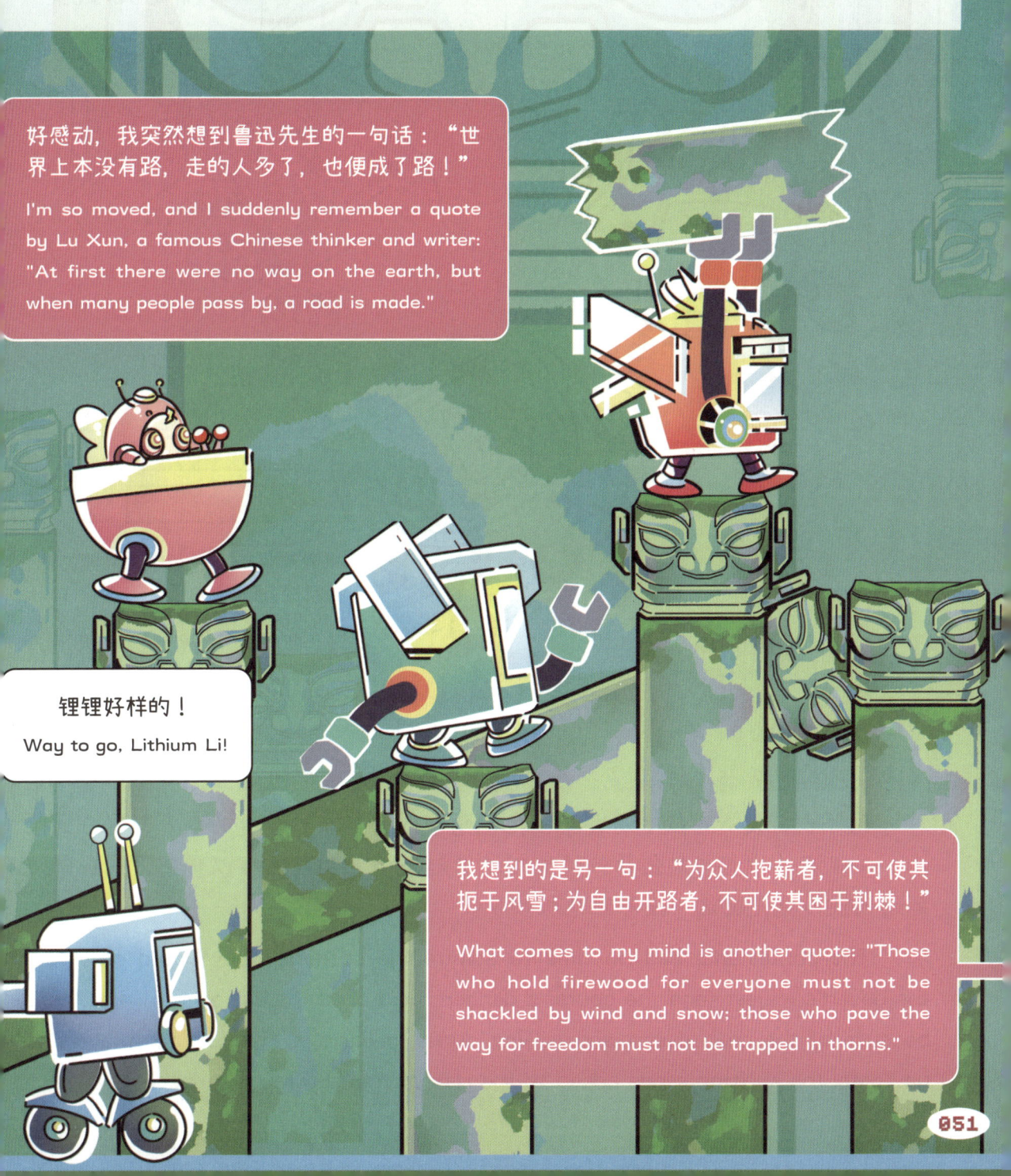

好感动，我突然想到鲁迅先生的一句话："世界上本没有路，走的人多了，也便成了路！"

I'm so moved, and I suddenly remember a quote by Lu Xun, a famous Chinese thinker and writer: "At first there were no way on the earth, but when many people pass by, a road is made."

锂锂好样的！

Way to go, Lithium Li!

我想到的是另一句："为众人抱薪者，不可使其扼于风雪；为自由开路者，不可使其困于荆棘！"

What comes to my mind is another quote: "Those who hold firewood for everyone must not be shackled by wind and snow; those who pave the way for freedom must not be trapped in thorns."

历经万般艰险，一众选手在锂锂的带领下，终于一起跨越过了障碍重重的石林路段，美丽的大漠景致映入眼帘。

Getting through numerous difficulties, the contestants led by Lithium Li finally crossed the obstacle-laden stone forest. The beautiful desert came into view.

太热了，45度的高温，我……我撑不下去了。
It's too hot with forty-five degrees Celsius! I... I can't hold on.

糟糕，拔不出来了。
Oh no, I'm stuck.

冲到最大功率……唉，筋疲力尽，我已经拼尽全力了。
Running at full power... Ugh, I'm exhausted. I've given it my all.

选手们可没心思欣赏这美景，他们感受到的只有大沙漠的威严与赛程的险恶——纵深的沙漠与起伏的沙梁使环境看起来四面相同，难以辨别方向；暗流涌动的流沙更是使这一段赛程危机四伏。

The contestants had no time to admire the scenery. They could only feel the desert's majesty and the course's danger—it looked the same in all directions because of the deep desert and rolling dunes, making it difficult for players to navigate; the shifting sands added to the risk of this section.

然而，除了地理因素，对电池选手们更要紧的一点是，此时温度竟高达45℃！在高温环境下，选手们不得不将机器人的动作放慢下来，以稳定自身的性能。

In addition to geographical factors, however, the greater challenge for the battery contestants was the temperature: forty-five degrees Celsius! In such heat, they had to slow down their robots' movements to stabilize their performance.

大家伙儿跟我来，前方有路标！

Everyone, follow me, there's a sign ahead!

小贴士 Tips

电池在高温下放电，会缩短其使用寿命。如果超过 45℃，会破坏电池内的电化学平衡，导致副反应发生。

When batteries are discharged at high temperatures, their lifespan will be shortened. If it exceeds 45 degrees Celsius, the electrochemical balance inside the battery will be disrupted and side reactions could be cuased.

难能可贵的是，在这样极限的考验当中，选手们没有再一味地追求比赛的胜利，反而开始互相帮助起来。

It is commendable that the contestants stopped blindly chasing victory in this extreme race but began to help each other.

加油！我们都要坚持住！胜利就在前方！
Come on! Hang in there! Victory is in the front!

你的问题不大！振作起来！
It's not a big deal. Hang in there!

谢谢你！
Thank you!

如果还有下次的话，建议提前准备充分一些！
If there's a next time, be more prepared!

别担心，温度降下来后很快就能恢复。
Don't worry, once the temperature drops, you'll recover.

不过，也不是所有电池人都能有这样的觉悟。瞧，镍霸又违反规则了，这次受害的是锰叨叨，她迷路之后好一阵才被工作人员领回去。

However, not all battery contestants shared the spirit. Look, Tyrant Nickel broke the rules again. This time Mnag the victim got lost and was brought back by the staff after a while.

看我弄乱方向标！

I'm going to mess up the direction signs.

小贴士　Tips

在所有的环境因素中，温度对电池的充放电性能影响最大。电池的心脏——电极和电解液——与环境温度的变化息息相关。

Among all environmental factors, temperature plays the most significant role in a battery's charging and discharging performance. The heart of the battery—the electrode and the electrolyte—are closely related to changes in ambient temperature.

裁判救我！我申请放弃比赛！我不想一个人待在这里！

Help, referee! I want to give up the race! I don't want to be here alone!

从高达 45 摄氏度的沙漠赛段出来，接着又投入零下 10 摄氏度的冰川赛段，对选手们来说，这无疑是一次"才出烤箱，又进冰箱"的极端考验。

After leaving the desert where the temperature reached forty-five degrees Celsius, the contestants entered the glacier section where it was minus ten degrees Celsius. It felt like "going from the oven to the freezer". What an extreme race!

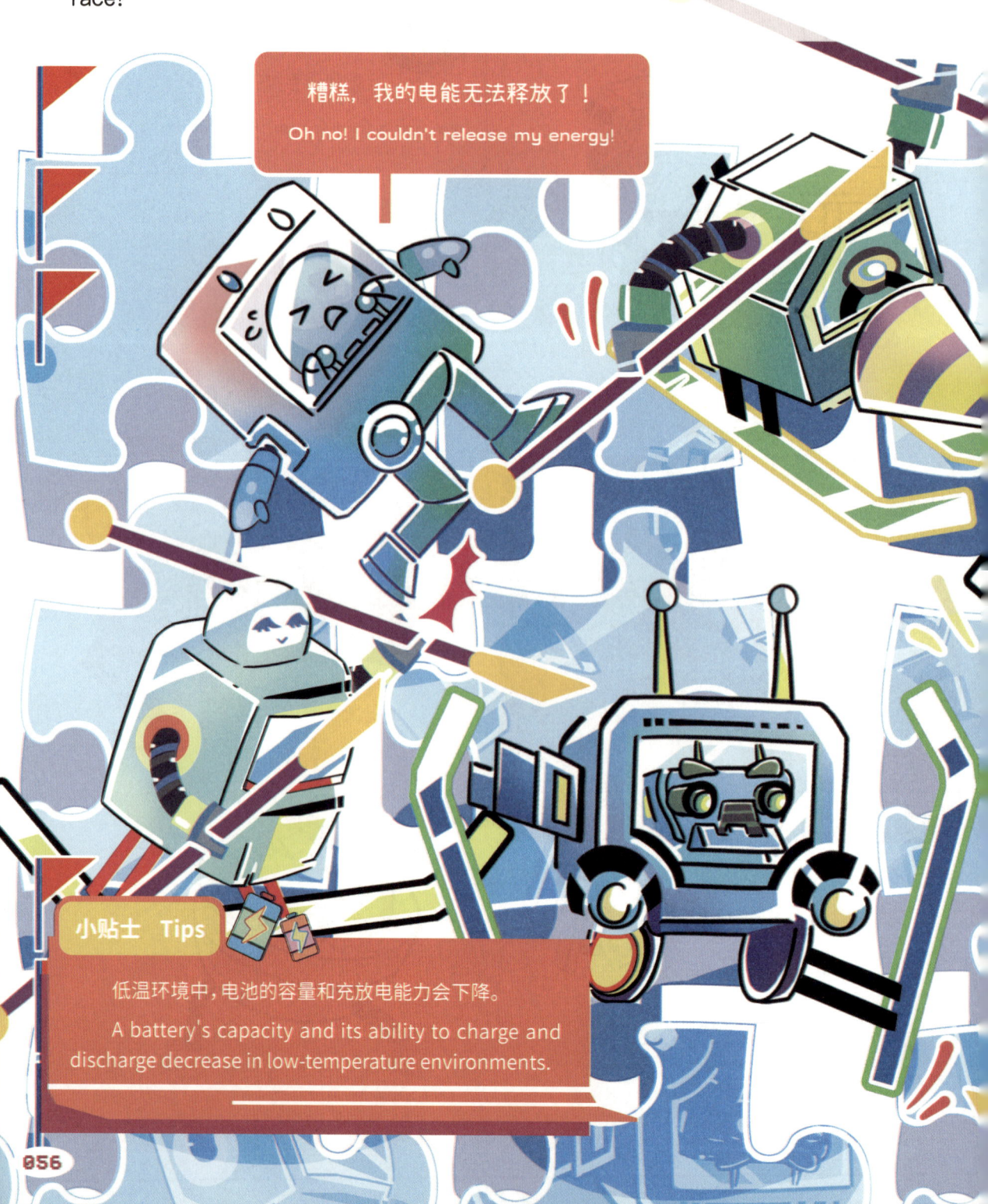

糟糕，我的电能无法释放了！
Oh no! I couldn't release my energy!

小贴士 Tips

低温环境中，电池的容量和充放电能力会下降。
A battery's capacity and its ability to charge and discharge decrease in low-temperature environments.

一片冰封的湖面迎面而来，出发时还浩浩荡荡的队伍此时只剩下了寥寥数人。早有准备的几位选手不慌不忙地将机器人切换到冰面滑行模式，无法改变形态的机器人只能在冰面上缓步前行，没多久就电量耗尽了。

A frozen lake appeared ahead. By this point, only a few contestants from the large group at the start remained. Those well-prepared contestants calmly switched their robots to ice-skating mode, while those with robots unable to change forms could only move slowly on the ice, soon running out of power.

啊……太冷了……
Ah… too cold…

眨眼间就迎来了冰川赛段最后一个考验，也是本次障碍越野赛的最后一个障碍——跨越湖面上漂浮的碎冰。

In the blink of an eye, the final test of the glacier section came. It's also the last obstacle of this race—crossing the floating ice on the lake.

稳住！越过这最后一个障碍就是公路了，我依然有优势！

Hold on! Once I get over this last obstacle, it will be a clear road ahead. I still have the advantage!

别担心！我们来帮你！

Don't worry! We can help you!

要以不失力量却又轻盈灵敏的动作起步，才能在空中完成转体动作，成功抵达彼岸。这样的动作对驾驶装置的要求极高，稍有不慎便会在中途跌落到冰冷的湖水之中。

It would take both strength and agility to complete a turn in the air and reach the other side. Such maneuvers demanded a lot for the device. A slight mistake could result in a fall into the icy water.

为了我的家族，我一定要赢！
I must win! For my family!

完了，进水了，动不了了。
Oh no, water, I can't move.

想不到，我氢天天英明一世……
It's unexpected. Me, Hydrogen, wise all my life...

罢了罢了，是金子，在哪里都会发光的。
Forget it. Gold shines everywhere.

小贴士 Tips

设备进水对电池的危害主要是可能导致电池短路，而电池短路后便无法再使用。

The main harm of water ingress into the device to the battery is the possibility of causing short circuit, rendering it unusable afterward.

锂锂驾驶着电动汽车率先到达终点，大铅、锌博士、镍霸驾驶的机器人也相继抵达，晋级为四强。

Lithium Li driving the electric car was the first to reach the finish line, and robots driven by Big Lead, Doctor Zinc, and Tyrant Nickel also arrived one after another, advancing to the top four.

细节决定成败！
Details determine success or failure!

终 Finish
点 Line

恭喜，晋级赛第一名产生了，是锂锂！

Congratulations! The first place in the advancement race goes to Lithium Li!

终点线　Finish Line

听说您的驾驶装置会变形，您能说说是怎么回事吗？

I heard your driving device can transform. Could you tell us more about it?

恭喜您获得第一名，您的最终目标就是冠军吗？

Congratulations on your victory! Is the championship your final goal?

非常感谢大家的关注，我想，现在我需要休息！

Thank you all for your attention, I think I need to have a rest now.

第一名锂锂，第二名大铅、第三名锌博士、第四名镍霸，恭喜四位实力强大的选手成功晋级！

First place: Lithium Li; Second place: Big Lead; Third place: Dr. Zinc; Fourth place: Tyrant Nickel. Congratulations to the four strong contestants for advancing!

这些头盔是为你们量身打造而成，它们不仅仅代表了你们不惧艰险、顽强拼搏的精神，更是我们电池王国的使命、荣誉、责任的象征！

These helmets are custom-made for you. They represent your courage and perseverance, symbolizing the mission, honor, and responsibility of our Battery Kingdom!

"金头盔"奖项无关于名次！在这场赛事当中，你们能够完赛，就是胜利！

The 'Golden Helmet' award isn't about ranking! In this event, finishing is victory!

4

艰难晋级！我有绝技！

Tough Advancement! I Have a Secret Skill!

本次大赛表现优异的选手俨然成为电池王国全民热议的对象，在大批粉丝的追捧下，大街上贴满了四人的海报，商店也纷纷推出了关于他们的周边产品。

The top performers in the competition became hot topics in the Battery Kingdom. With supports from tons of fans, posters of the four finalists were everywhere on the streets, and shops were rolling out products featuring them.

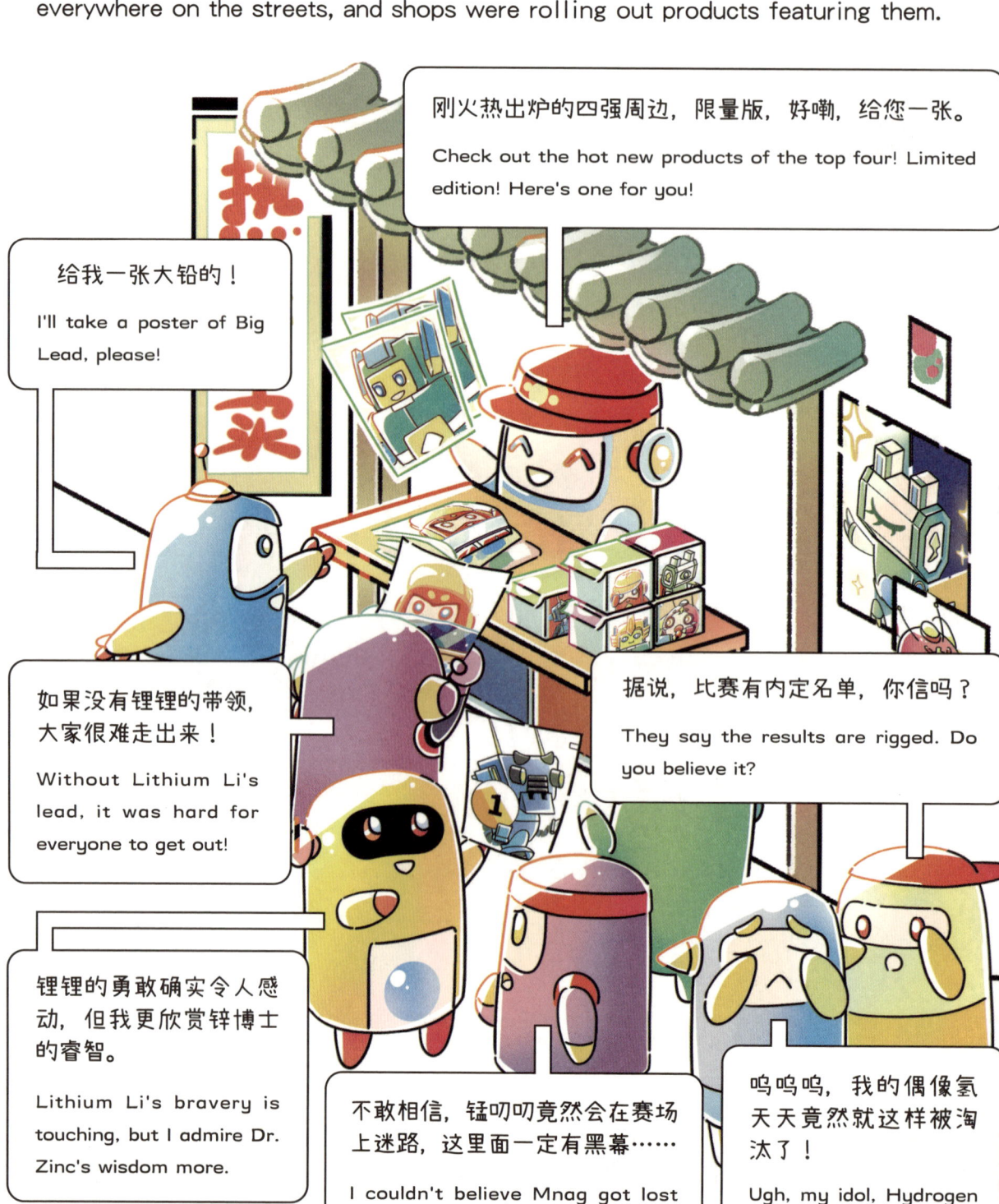

刚火热出炉的四强周边，限量版，好嘞，给您一张。

Check out the hot new products of the top four! Limited edition! Here's one for you!

给我一张大铅的！

I'll take a poster of Big Lead, please!

如果没有锂锂的带领，大家很难走出来！

Without Lithium Li's lead, it was hard for everyone to get out!

据说，比赛有内定名单，你信吗？

They say the results are rigged. Do you believe it?

锂锂的勇敢确实令人感动，但我更欣赏锌博士的睿智。

Lithium Li's bravery is touching, but I admire Dr. Zinc's wisdom more.

不敢相信，锰叨叨竟然会在赛场上迷路，这里面一定有黑幕……

I couldn't believe Mnag got lost in the game. There must be something shady going on…

呜呜呜，我的偶像氢天天竟然就这样被淘汰了！

Ugh, my idol, Hydrogen Hero, got eliminated just like that!

万众瞩目的晋级赛结束了，四强名单出炉，决赛冠军会是谁呢？人民们对于比赛的猜测也愈演愈烈，众说纷纭。

The highly anticipated advancement race had ended, and the list of the Top Four had been revealed. But who would be the champion of the final? The nation was buzzing with guesses, and everyone had his/her own opinion.

今天的饮料免费请二位勇士喝！
Drinks are free for you two, brave warriors!

听说这届电池竞技大赛是为国王选拔勇士。
I heard this competition is a selection for the King's Warrior!

还说要进亲卫队呢！
And the winners will join the royal escort!

嘿嘿，我还挺喜欢这个称号的！
Haha, I quite like this title!

勇士？这些都是谣言吧！
Warrior? That's only rumor, right?

在紧凑的赛程安排下，锂锂、大铅、锌博士、镍霸都怀着紧张而又激动的心情来到了四进二比赛现场，选手需要通过现场抽签确定对手，进行两两对决。

With a tight schedule ahead, Lithium Li, Big Lead, Dr. Zinc, and Tyrant Nickel arrived at the four-to-two semifinals. They were nervous but excited. Contestants drew lots for their matchups for the two-versus-two battles.

锂锂抽到的是1号，对手是抽到3号的锌博士；抽到2号的大铅对战的是抽到4号的镍霸。

Lithium Li, number 1, faced Dr. Zinc, number 3; Big Lead, number 2, took on Tyrant Nickel, number 4.

还记得我们的约定吗？
Do you still remember our promise?

当然记得！决赛见！
Of course I do! See you in the final!

四进二比赛是在虚拟空间中进行的，选手们可以充分发挥自己的想象力，用电力幻化出自己需要的一切。

平日里，锂锂与锌博士在驾驶电动汽车的工作中就常常被拿来作比较，这还是头一次正面切磋呢，两人已经迫不及待了。

The semifinals took place in a virtual arena where contestants could fully unleash their imaginations, using electricity to conjure anything they needed.

In their daily life, Lithium Li and Dr. Zinc were often compared in powering electric cars. It was the first time that they faced off each other. They couldn't wait to start.

> 这次的比赛项目需要选手们戴上VR头显，在虚拟空间中进行电力比拼。
>
> The race requires the contestants to wear VR headsets and engage in a power showdown in the virtual space.

> 四进二第一场，锂锂对战锌博士，开始！
>
> Semifinal round 1: Lithium Li versus Dr. Zinc! Begin!

> 冠军是我的！
>
> The championship is mine!

让我们来比试一下！
Time for a showdown!

天哪，绝技表演啊！
Wow, stunt performances.

我能将空气转化为我的能量！
Watch me turn air into energy!

哇哦！
wow!

小贴士 Tips

锌—空气电池的正极活性物质来自于电池外部的空气中所含的氧，理论上有无限容量。

The positive active material in zinc-air batteries comes from the oxygen in the air outside the battery. Theoretically, it has unlimited capacity.

你确实很强，但我也不差！
You're amazing, but I've got my own strengths!

我想，你还差点儿想象力！
I think you're a bit short of imagination!

在这场电力比拼中，锂锂发挥了十足的想象力，既赢得了胜利，还给观众们带来一场视觉上的新奇体验，不过不知怎么了，比赛一结束他竟晕倒过去。

In this electricity race, Lithium Li used his imagination to the fullest and not only won the victory, but also brought novel visual experience to the audience. However, for some reason, he fainted as soon as the race ended.

四进二第一场，锂锂对战锌博士，锂锂胜出，成功晋级决赛！
Semifinal round 1: Lithium Li versus Dr. Zinc, Li won and advances to the final!

喂！你没事儿吧？
Hey! Are you okay?

耶！太……
Yeah! So...

锂锂怎么了？
What happened to Lithium Li?

我早说了，安全最重要！
I always said safety comes first!

在赛场上，即使是深受人们喜爱的选手遇到了意料之外的状况，比赛仍然要进行下去，这就是规则。

第二场大铅与镍霸的比拼如期进行在即，观众们在裁判的示意下渐渐安静下来。

On the race course, even when a beloved contestant encountered unexpected issues, the game carried on—that's the rule.

The second match, between Big Lead and Tyrant Nickel, was about to proceed as scheduled. The audience quieted down at the referee's signal.

大家安静！立即送锂锂到医院救治。2号对4号比赛继续进行。

Everyone, calm down! Send Lithium Li to the hospital immediately. The match between No. 2 and No. 4 continues.

马上到大铅了，我要接着看完！

Big Lead is up next. I must watch this!

我要去看看锂锂怎么了！

I need to check on Lithium Li.

锂锂，你怎么了？

Lithium Li, are you okay?

四进二第二场，大铅对战镍霸，开始！

Semifinal round 2: Big Lead versus Tyrant Nickel! Begin!

大铅和镍霸的决斗正式开始，虚拟空间却莫名变得很不稳定，观众们只看得到卡住的画面，便纷纷抗议起来。

镍霸趁机给大铅说了许多没头没尾的话，大铅只当他又在耍花招，很快便进入了状态。

As Big Lead and Tyrant Nickel began their duel, the virtual space unexpectedly became unstable. The audience could only see a frozen screen and began protesting loudly.

Tyrant Nickel took the chance to ramble on to Big Lead with a bunch of nonsense. Big Lead dismissed it as another trick and quickly got back into focus.

嘿，兄弟，先别动手！
Hey, buddy, hold on a second!

你愿意加入我们吗？
Would you like to join us?

这次你又要耍什么花招！
What trick are you up to this time?

各位观众稍安勿躁，设备系统出现故障，正在紧急维护中。
Attention, audience. Please stay calm. There are technical difficulties in the system and it is under urgent maintenance.

不是吧？刚刚才坏了个电池，这下机器也坏了？
No way! A battery just went down, and now the machines?

这让我们看什么？
What are we supposed to watch now?

卡了？ Stuck?

听着，兄弟，我很欣赏你。我不应该是你的敌人，我们只有联合起来……
Listen, bro, I admire you. I shouldn't be your enemy; we need to unite...

没时间了！你记住，我的现在，就是你的未来！
There's no time! Remember, my present is your future!

别废话了！接招吧！
Enough talk! Take it!

075

毫无疑问，专心战斗的大铅击败了镍霸，实现了"与锂锂决赛相见"的约定。

Undoubtedly, the focused Big Lead defeated Tyrant Nickel, fulfilling the promise "to meet Lithium Li in the final".

我觉得工作人员应该出来给个说法。
I think the staff should come out and give an explanation.

啥也没看到！没劲！
I didn't see anything! How boring!

就这样结束了吗？
Is it over just like that?

看着台下熙熙攘攘，镍霸只是一脸颓然地站在竞技台上。

Seeing the bustling crowd below, Tyrant Nickel stood on the stage, discouraged.

四进二第二场，大铅对战镍霸，大铅胜出，成功晋级决赛！
Semifinal round 2: Big Lead versus Tyrant Nickel, Big Lead won and advances to the final!

5

新王诞生

Rise of the New King

四进二比赛一结束，大铅就火急火燎地赶到了医院看望锂锂，原来锂锂是因为电能消耗过大，体力不支晕过去了。

As soon as the semifinal ended, Big Lead rushed to the hospital to see Lithium Li. It turned out that Lithium Li fainted for using too much energy and getting tired.

醒了，醒了！
He is awake! He is awake!

我还好，比赛进行得如何？
I'm okay. How did the match go?

太好了，你现在觉得怎么样？
Thank goodness! How are you feeling now?

镍霸输了，你和我进入决赛了！
Tyrant Nickel lost. You and I are in the final!

这么说，我们的约定实现了？！
So, we keep our promise?!

你这次晕倒，是由于过度放电导致的。
You fainted due to over-discharge.

不受保护的过度充电和过度放电，将对你的正负极造成永久的损坏。
Unprotected overcharging and over-discharging can damage your electrodes for good.

为了你的身体健康，即使是比赛，也不要再逞强。
For your health, don't push yourself too hard, even in a race.

锂锂刚从昏迷中醒来，就关心起比赛的情况，它担心镍霸再使坏而伤害大铅。得知镍霸怪异的行为后，虽然感到不解，但最终决赛在即，锂锂也没有时间多想了。

Lithium Li was just awake from a coma and immediately began to concern about the race. He feared Tyrant Nickel might harm Big Lead. Even though Lithium Li didn't understand why Tyrant Nickel was acting so weird, there was no time to think about it. The final was coming soon.

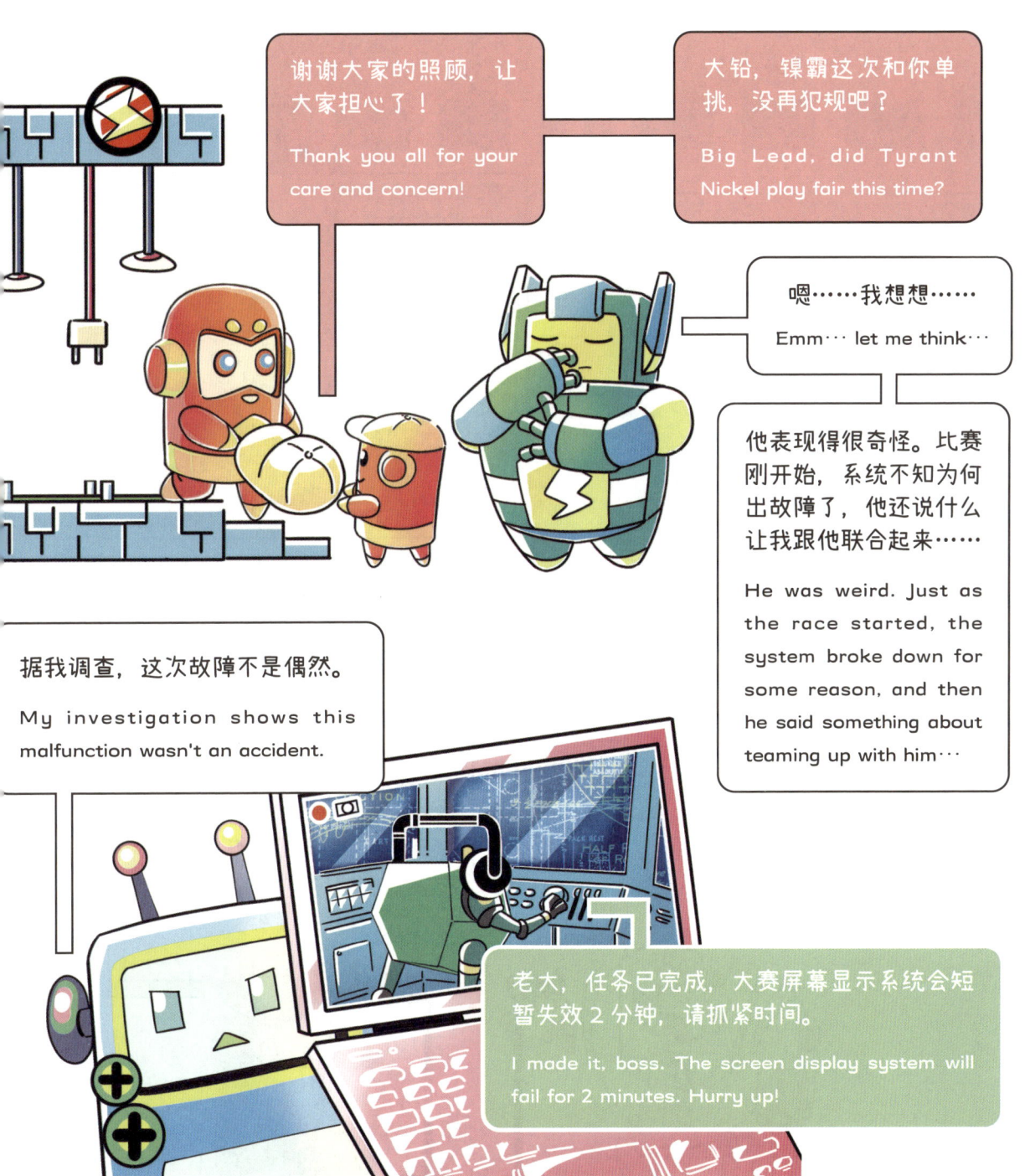

谢谢大家的照顾，让大家担心了！

Thank you all for your care and concern!

大铅，镍霸这次和你单挑，没再犯规吧？

Big Lead, did Tyrant Nickel play fair this time?

嗯……我想想……

Emm… let me think…

他表现得很奇怪。比赛刚开始，系统不知为何出故障了，他还说什么让我跟他联合起来……

He was weird. Just as the race started, the system broke down for some reason, and then he said something about teaming up with him…

据我调查，这次故障不是偶然。

My investigation shows this malfunction wasn't an accident.

老大，任务已完成，大赛屏幕显示系统会短暂失效2分钟，请抓紧时间。

I made it, boss. The screen display system will fail for 2 minutes. Hurry up!

次日，锂锂与大铅的对决即将拉开帷幕，国王也来到了比赛现场，宣布了他将亲自考核总冠军的重磅消息，这让观众们更加议论纷纷。

The next day, the showdown between Lithium Li and Big Lead was about to begin. The king arrived at the venue, announcing he would personally evaluate the champion, which caused more discussion among the audience.

电池竞技大赛——巅峰总决赛
Battery Competition—— The Final

姓名：__王一硫__
种族：__二次电池__
家族：__钠-硫电池__
适应温度：__300～350 ℃__
主要应用领域：__削峰填谷、应急电源、风力发电、储能等。__

Name: __King Monosulfide__
Type: __Secondary Battery__
Family: __Sodium-sulfur Battery__
Operating temperature: __300~350 ℃__
Main application areas: __Balancing electricity supply, emergency power, wind power, energy storage, etc.__

我们电池面临的考验，不仅仅是容量、功率、安全性等方面，对环境的适应性也日益重要。

As batteries, we face many challenges. We need to store a lot of energy, be powerful, and be safe. Besides, it is important for us to adapt to various environments.

决赛将会重点对此方面进行考核，希望两位选手不负众望，比出精神，拼出实力！最后，我将会亲自考核这届的冠军！

The final will focus on these aspects. We hope the two contestants will live up to expectations, demonstrate their spirit, and put in their strength! Finally, I will personally evaluate the champion of the year!

最终决赛是水上救援大比拼，选手们需要驾驶救生船，在规定时间内尽可能多地救援漂浮在河里的电池人。

怕水几乎是每一块电池的弱点，这场比赛，既是对选手们综合能力的考核，也是对其心理素质的考验。

The final was a water rescue race. Contestants must steer lifeboats and save as many floating batteries as possible within the time limit.

Almost every battery is afraid of water. This race tested both the comprehensive skills and psychological quality of contestants.

水上救援大比拼
Water Rescue Race

在20分钟内，救下1个电池得1分，累计得分高者胜出。

In 20 minutes, saving 1 battery earns 1 point. The one with highest points wins.

提示：电池漂浮最长时间为 20 分钟，30 分钟后将沉下去。

Note: Batteries can float for 20 minutes at most. After 30 minutes, they will sink.

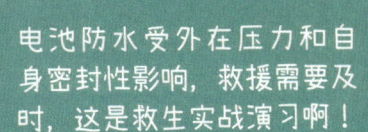

电池防水受外在压力和自身密封性影响，救援需要及时，这是救生实战演习啊！

Battery waterproofing is affected by external pressure and its own sealing. Quick action is required for rescue—this is a real-life rescue drill!

说实话，我非常怕水。

To be honest, I'm very afraid of water.

我也怕，不过不用担心，如果发生意外的话，裁判会救援我们的。

Me too, but don't worry. If anything happens, the referee will rescue us.

比赛刚开始,天空忽然变得乌云密布,一道道闪电划破长空。

As the race began, the sky suddenly darkened, and lightning streaked across the sky.

两人不分伯仲啊!
They're neck and neck!

锂锂
Lithium Li
11

大铅
Big Lead
10

在天气的变化下，赛事显得愈发焦灼起来。锂锂目前领先大铅1分，大铅心急如焚，竟不顾救生船仅限1人的载量，同时打捞起2个电池人往回走，这使它的救生船不断晃动，显得很不稳定。

Under the changing weather, the race became more intense. Lithium Li led Big Lead by 1 point. Big Lead was so anxious that he disregarded the limited capacity of one person on the lifeboat and attempted to rescue 2 batteries at the same time. Then, he started to head back. However, overloading caused the lifeboat to sway uncontrollably and appear unstable.

不好，大铅的操作好危险！
Oh no, It's dangerous!

天气突变，请选手们立即上岸！比赛应以安全作为前提！

The weather has changed! Contestants, return to shore now—safety comes first!

超载并没有使大铅拉回得分，反而使水渗入了救生船内部，救生船向下沉了一大截。大铅的身体也沾到了水，浑身打颤起来，他只得忍着身体的不适坚持操作，最终只将一个电池人拖上了岸边，眼睁睁看着另一个电池人随水流飘走了。锂锂见大铅正在下沉，果断向铅大丢出了救生绳，这才将他拉了回来。

Overloading did not help Big Lead gain points. Instead, it caused water to seep into his lifeboat, causing it to sink a lot. Big Lead's body also got wet, and he started to shiver. He had to endure the discomfort and kept operating. In the end, he only managed to drag one battery to the shore, and watched the other drift away with the current. Seeing that Big Lead was sinking, Lithium Li threw a rescue rope to him and pulled him back to safety.

时间到，比赛结束。
Time's up, the race is over.

快，抓住救生绳！
Hurry up, grab the rope!

安全最重要！
Safety first!

谢谢你！我为了追回得分，急功近利，反而给自己带来了危险！
Thank you! I was so eager to win, only to put myself in danger!

分数统计
Score

12　　11

巅峰对决，锂锂胜出，即将接受国王的考验！
A big showdown! Lithium Li won! He is going to take the king's test!

比赛真精彩，刚开始我以为大铅会赢！
What a thrilling race! I thought Big Lead would win at first!

你们看到没？锂锂救了大铅！
Did you see? Lithium Li saved Big Lead!

是呀，如果大铅最后能将那两个电池人拖上岸，他就赢了！
Yeah, if he had dragged the two batteries to the shore, he would have won!

观众们跟随比赛转移到了电池王国等级最高的比赛场馆——"能量之冠"场馆，来到了"王之挑战"比赛现场。

大家即将见证电池王国历史上最激动人心的挑战，一个是实力深不可测的国王，一个是从众电池中脱颖而出的新秀，谁输谁赢，以及国王发起挑战的原因，都成了国民们口中津津乐道的谈资。

The audience moved to the highest-level arena in the Battery Kingdom—the "Crown of Energy"—to witness the "King's Test".

They were going to witness the most exciting challenge in the Battery Kingdom's history. A strong king versus a rising battery star. Who would win? Why did the king launch this challenge? These were hot topics among the nation.

众所周知，我们国王英明神武，治国有方，不然我们电池王国也不会发展这么快，他聪明又能干，公正又无私……

Our king is wise and capable, well-known for his excellent governance. Otherwise, our Battery Kingdom wouldn't have developed so quickly. He is smart, capable, fair, and selfless…

不同的电池大家族虽然个性迥异，但是却能和睦相处，正是因为我们拥有这样一位伟大的国王。你们说，锂锂靠什么能赢呢？

Although each battery family has unique personality, we all live in harmony, because we have our great king. So, what do you think— what will make Lithium Li stand a chance to win?

冠军授奖仪式结束，有请尊敬的国王——王一硫！
The championship award ceremony has ended. Please welcome our respected king—King Monosulfide!

各位电池王国的公民们，锂锂已通过了大赛的各项考核。现在，我也将兑现自己的承诺——亲自对他进行考验。
Nation of the Battery Kingdom, Lithium Li has passed all competition tests. Now, I will fulfill my promise—to test him in person.

这是一场王位挑战赛，年轻人，你做好准备了吗？
It is a challenge for the throne. Young man, are you ready?

请国王赐教！
Yes, my King!

你说挑战国王意味着什么呢？
What does challenging the king mean?

如果赢了能当国王？如果输了，不好说！
The winner can become the king? And what happens to the loser? Who knows!

对于王之挑战赛，国王安排的是最简单直接的比赛方式：驾驶战斗机器人，每击中对手的机器人一次获得1分，三局两胜。

比赛刚刚开始，国王驾驶机器人猛地发射出一束束激光，锂锂环顾四周，避无可避，被击中后直接摔倒在地，丢掉1分。

The King's Test has a very simple and direct rule: the contestant drives a combat robot and earns 1 point for each hit on the opponent's robot, and the one who earns two points first wins in three rounds.

The battle had just begun, and the king's robot fired a series of lasers. Lithium Li looked around and found no way to dodge. He was hit and fell to the ground. Lithium Li lost 1 point.

你不知道吧，国王所属的家族是钠-硫电池家族，他们可以大电流、高功率放电。瞬时间可放出3倍固有能量，可谓无人能敌！

You may not know that the king's family, the Sodium-sulfur Battery, is capable of discharging large current and high power. They can release three times their inherent energy in an instant. Unbeatable!

在绝对实力面前，锂锂遭到了压制！
Lithium Li is suppressed by absolute strength!

正当锂锂面对国王的火力强攻而一筹莫展之时，他忽然发现国王从比赛开始就不曾移动过自身位置。他这才想起来机器人X曾告诉过他，钠-硫电池家族在工作时不擅长移动，灵活性与安全性不够好！

想到这一点，锂锂重拾信心，利用自身的轻巧在国王身边不断游走，一边躲避激光，一边寻找着机会。

Just as Lithium Li felt stuck against the king's strong attack, he realized that the king hadn't moved his position since the start of the battle! Then, he remembered what Robot X told him. The sodium-sulfur battery family isn't good at moving during work, and their flexibility and safety were not good enough.

This gave Lithium Li confidence again. He used his lightness to constantly move around the king, avoiding the laser while searching for an opportunity.

国王为了定位高速移动的锂锂，不得不将攻势放缓，锂锂趁此间隙，凌空一跃，从空中发起攻击，一击即中。

The king had to slow down his attack to track the fast-moving Lithium Li. Seizing the moment, Lithium Li leaped into the air and launched an attack, hitting the target with one strike.

双方平分了，稍事歇息，最后一局开始了。面对锂锂以守为攻、以退为进的战术，国王显得应接不暇。

　　但此时此刻，他更加确信自己没有看错——能够引领电池王国各个家族走向更加繁荣道路的领袖，一定要具备沉几观变和不断追求进步的精神。

　　锂锂再次找准机会，发起了最后一击，以2：1的成绩打败了国王，赢得了最后的胜利。

The two tied. After a short break, the final round began. Facing Lithium Li's defensive tactics, the king seemed overwhelmed.

But at this moment, the king was even more certain that—— a good leader needs to stay calm, be observant, and keep pursuing better. Only with those qualities can he/she lead all the families toward prosperity.

Lithium Li seized the opportunity again and launched the final attack. He defeated the king with a score of 2:1 and won the final round.

哇！幻影移动！
Wow! A phantom move!

虽然我输了，但看到锂锂在比赛中的表现，我倍感欣慰。现在，电池王国正面临着巨大的危机，我年纪大了，无法在这场危机当中从容应对……

Although I lost the match, I am happy to see Lithium Li's performance. Our Kingdom faces a huge crisis now. I'm too old to handle it…

我宣布——我将正式禅位给锂锂。我们的国家将迎来一位更加勇敢、坚韧的国王！我相信他有能力保护每一位公民，为大家带来幸福。

I announce—I will pass on the throne to Lithium Li. Our kingdom will have a very brave and strong king! I believe he can protect you and bring happiness to everyone.

你要带领全国各个电池家族团结一致，共度危机！这是一项使命，也是一种责任，只有一代接一代的电池人不断努力，我们电池大家族才会持续昌盛！

You must bring all the battery families together to overcome the crisis! It's your mission! It's your responsibility. Only through the efforts of one generation after another can our big battery family continue to thrive!

危机？ Crisis?

他有什么能力带来幸福？我的家族每年都有不幸发生，谁能改变这一切！

What can he do to bring happiness? My family experiences misfortunes every year. Who can change this?

当锂锂走出赛场，门口早已被记者们围得水泄不通。对于国王提到的"危机"，电池王国人心惶惶。对此，锂锂十分不解，但同时这也让他意识到——在继任王位后，有更多问题等待他解决！

When Lithium Li walked out of the venue, he was surrounded by reporters. Everyone was anxious about the "crisis" the king mentioned. Lithium Li was confused, but he realized that—there would be many problems to deal with for a new king!

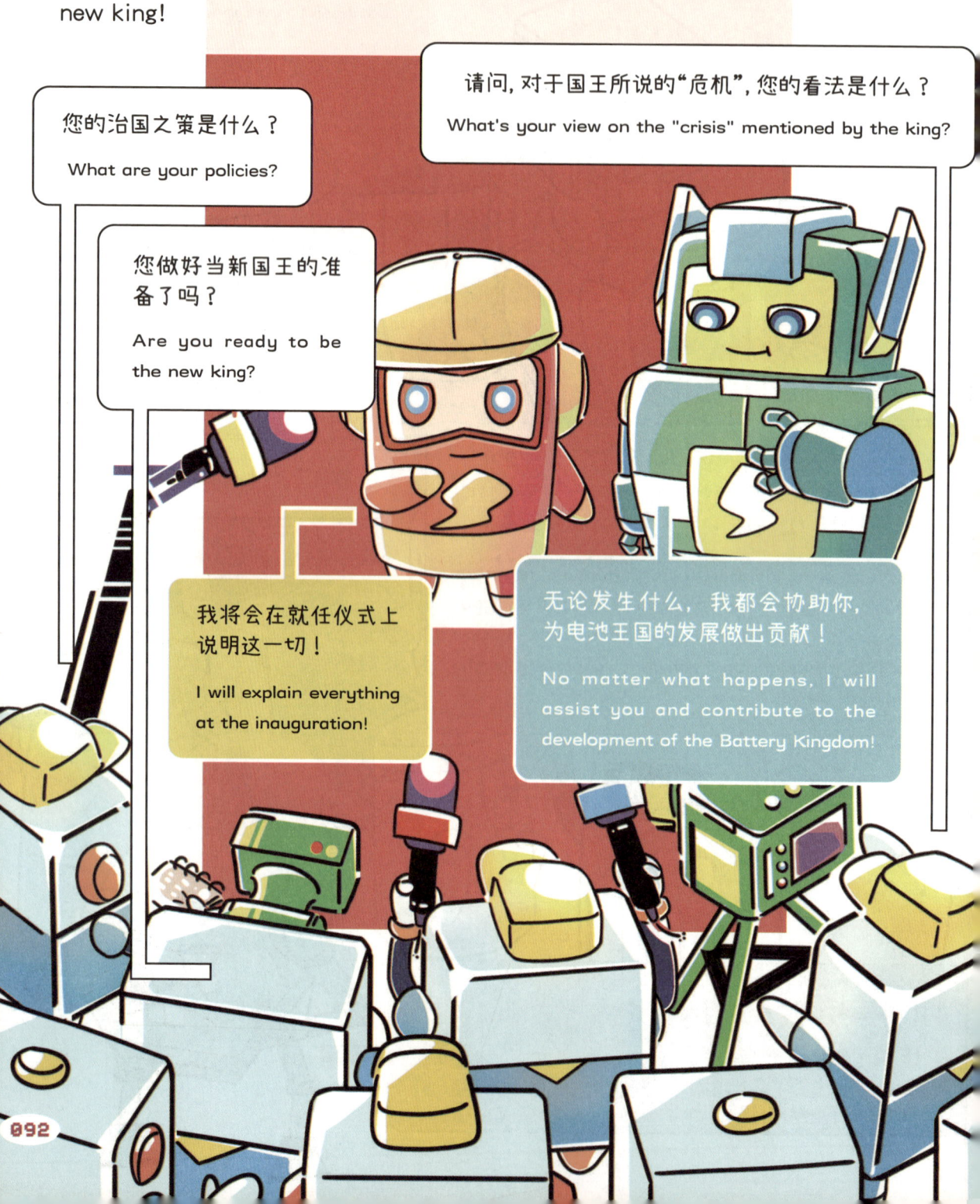

请问，对于国王所说的"危机"，您的看法是什么？
What's your view on the "crisis" mentioned by the king?

您的治国之策是什么？
What are your policies?

您做好当新国王的准备了吗？
Are you ready to be the new king?

我将会在就任仪式上说明这一切！
I will explain everything at the inauguration!

无论发生什么，我都会协助你，为电池王国的发展做出贡献！
No matter what happens, I will assist you and contribute to the development of the Battery Kingdom!

电池大揭秘

Secrets behind Batteries

电池的定义
Definition of Batteries

在现代社会生活中,电池无处不在。

In modern society, batteries are everywhere.

电池能驱动一辆汽车,这么大的能量是从哪里来的呢?

Batteries can power a car—where does all the energy come from?

电池内部的能量来源于物质的化学反应,这些反应导致正负极之间产生电势差,从而使能量以电的形式释放出来。

The energy inside a battery comes from chemical reactions. These reactions cause a difference in electrons between the positive and negative electrodes, releasing energy as electricity.

电池作为重要的电能来源，具有可以提供稳定电压、长时间稳定供电、受外界影响很小等特点。同时，电池结构简单，携带方便，充放电操作简便易行，不易受外界气候和温度的影响，性能稳定可靠。

Batteries are an important source of electrical power. They can provide a stable voltage, offer long-lasting power, and are not easily affected by the environment. Batteries have a simple structure and are easy to carry, charge and use. They remain reliable regardless of weather or temperature.

电池的工作原理
The Working Principle of Batteries

每一个电池都是一个能量转换器，化学能直接转化为电能是电池内部自发进行氧化、还原等化学反应的结果。

Each battery is an energy converter. Inside a battery, chemical energy is changed into electrical energy through spontaneous chemical reactions like oxidation and reduction.

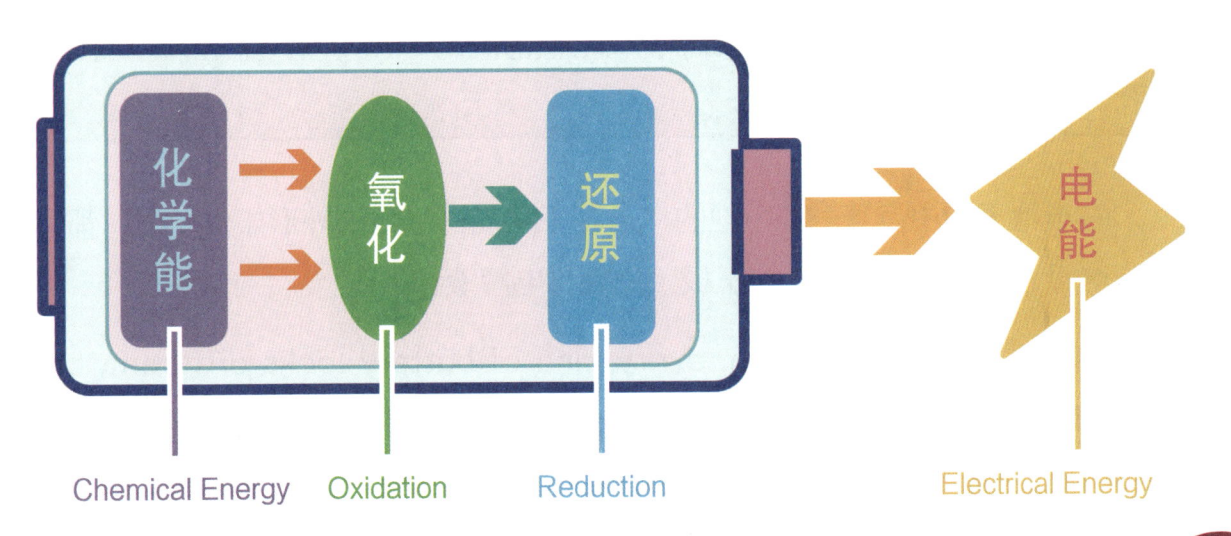

Chemical Energy — Oxidation — Reduction — Electrical Energy

要想真正理解电池的工作原理，让我们先来了解"水果电池"。

To understand how batteries work, let's first look at a "fruit battery".

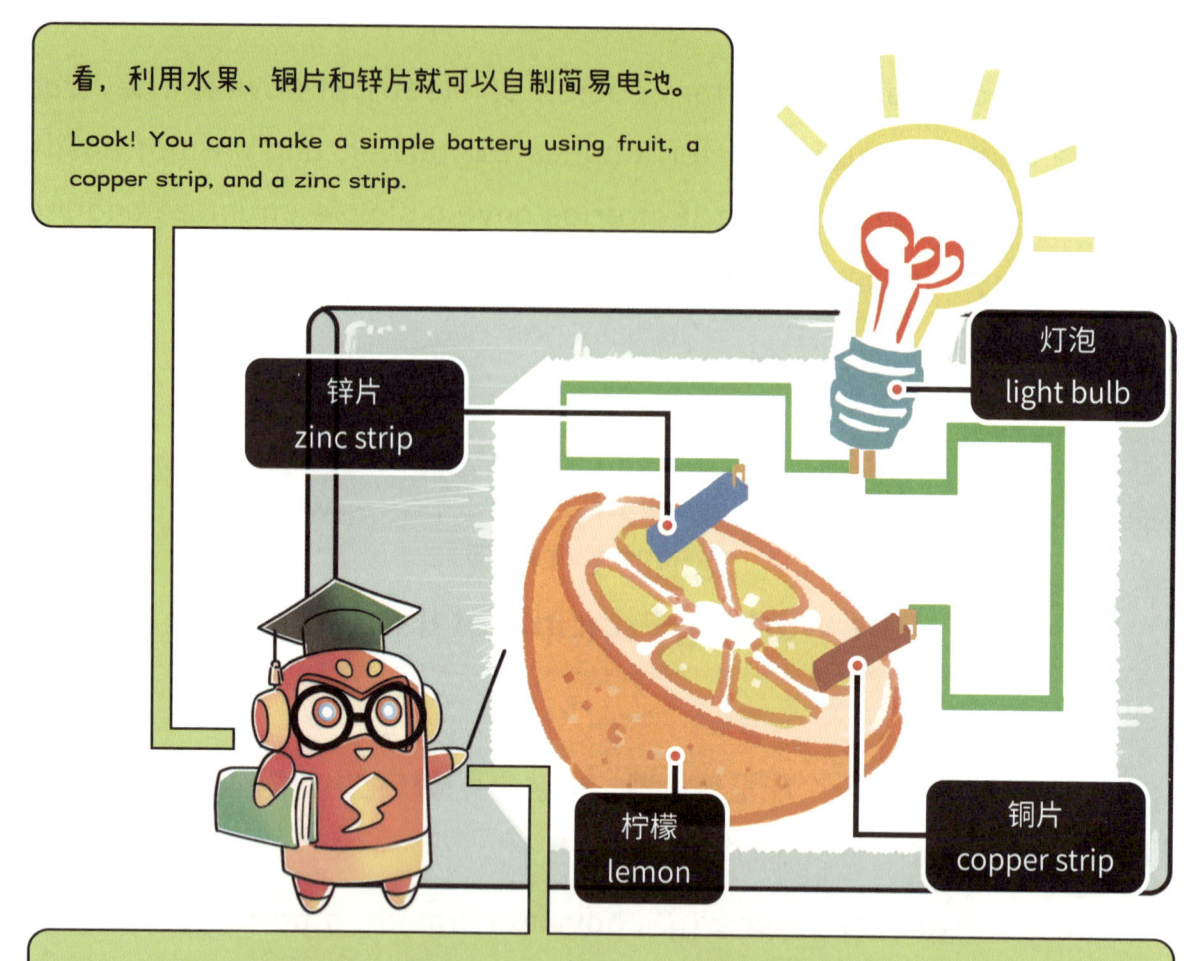

看，利用水果、铜片和锌片就可以自制简易电池。

Look! You can make a simple battery using fruit, a copper strip, and a zinc strip.

锌片
zinc strip

灯泡
light bulb

柠檬
lemon

铜片
copper strip

这里是以锌片、铜片作为电极材料，柠檬汁的生物酸作为电解液，并用导线将它们连接起来，形成闭合回路，这样就会有电子转移，灯泡就亮起来啦！

Here, the zinc and copper strips act as the electrodes, and the lemon juice works as the electrolyte. When you connect them with wires to form a circuit, electrons start to move, and the light bulb is lighted up!

在电池王国中，每一种电池的样子各不相同，但无论外形如何变化，都离不开这四大关键组成部分——正极、负极、电解液和外壳。

In the Battery Kingdom, every kind of battery looks different, but no matter what shape it is, all of them have four key parts: the positive electrode, the negative electrode, the electrolyte, and the casing.

电池为什么叫电"池"？
Why is it called "电池" in Chinese?

在将电池中的化学能转化为电能的过程中，电解液起到了至关重要的作用，导电以及化学反应都离不开它。

因此，电解液被人们誉为"电池的血液"。

In the process of turning chemical energy into electrical energy, the electrolyte plays an important role. It helps with both conducting electricity and chemical reactions. That's why people often call the electrolyte "the blood of battery".

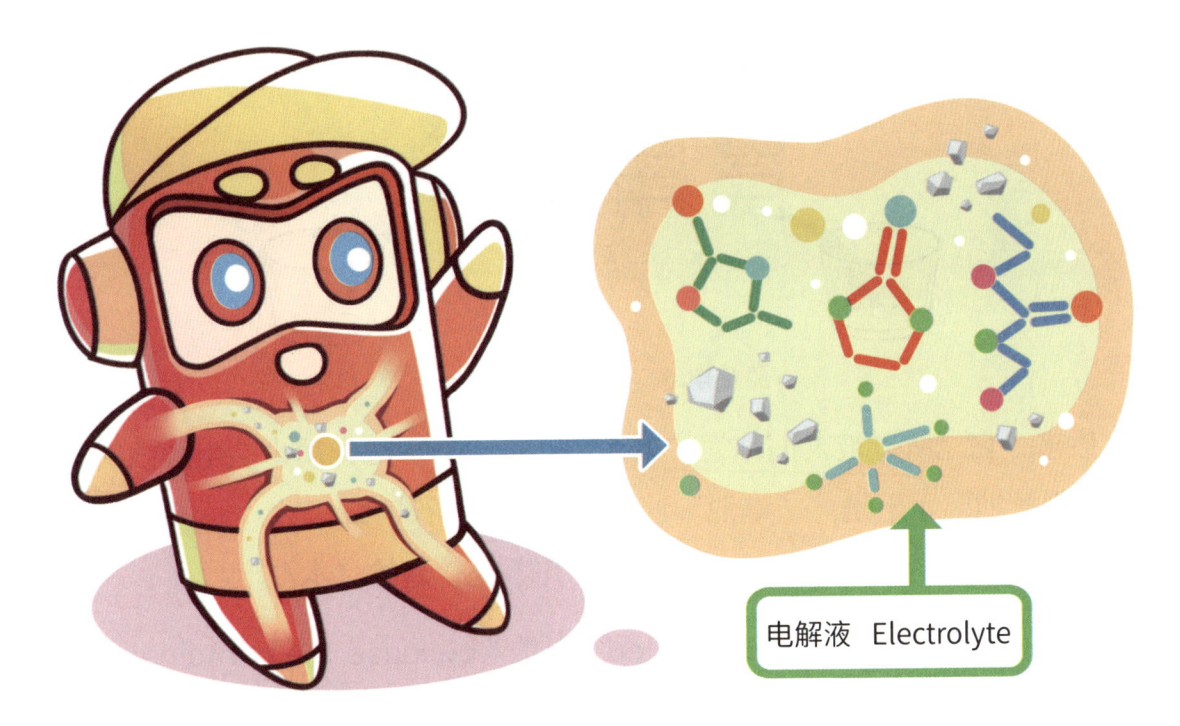

电解液 Electrolyte

那么"电池"这个名字到底从何而来呢？电池的"池"又是什么含义呢？

But where does the name "电池"(diàn chí) come from? What does the word "池" (chí) mean?

其实，在最早的时候，电池并不叫作"电池"，而是被称为"电堆""伏特堆"。

Actually, batteries weren't called "电池" in the very beginning. They used to be called "电堆" (electric piles) or "伏特堆" (volta piles).

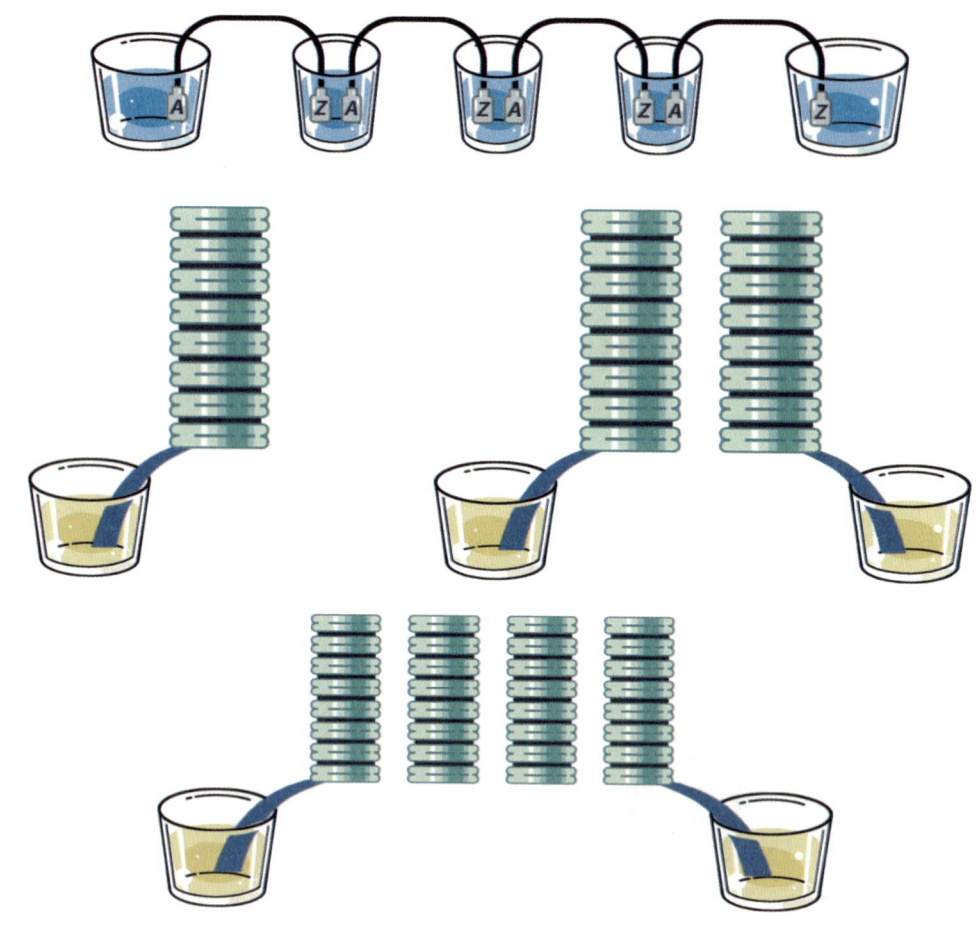

原始电池　Original Battery

后来，它之所以被人们称为"电池"，是因为那时候用来装电解液的容器叫作"电解池"。于是，"电池"这个名字就这样被叫开了。

所以，电池的"池"指的是电解池，并不是普通的"水池"。

Later, they started being called "电池" because the containers used to hold the electrolyte were called "电解池". That's how the name "电池" came about.

So, the "池" (chí) refers to the electrolyte cell, not a regular "water pool".

电池的分类
Types of Batteries

下图中这些形态各异的电池，都是现在国内外市场上的主流电池，那么我们该如何判断，它们在电池王国内各自属于哪一个派别的呢？

The various batteries shown in the picture below are all popular types in the domestic and international markets today. So, how can we tell which "family" each of these batteries belongs to in the Battery Kingdom?

锂离子电池	铅酸蓄电池	碱锰电池	镍镉电池
Lithium-ion Battery	Lead-acid Battery	Alkaline Manganese Battery	Nickel-cadmium Battery

镍氢电池	钠－氯化镍电池	钠－硫电池	锌－空气电池
Nickel-metal Hydride Battery	Sodium-nickel Chloride Battery	Sodium-sulfur Battery	Zinc-air Battery

电池种类繁多，遵循不同的前提条件，电池分类的方式有许多种。

电池按照能量转化的方式划分，可分为物理电池和化学电池；按电解液的种类划分，可分为酸性电池、碱性电池等；按正、负极所使用的材料划分，可分为锌锰电池、镍镉电池和锂电池等；按照工作性质的不同划分，可分为一次电池和二次电池……

There are many types of batteries, and there are different ways to classify them based on certain conditions.

Batteries can be divided into two main types based on how they convert energy: physical batteries and chemical batteries. They can also be classified by the type of electrolyte, such as acidic batteries, alkaline batteries, and so on. Depending on the materials used for the positive and negative electrodes, there are zinc-manganese batteries, nickel-cadmium batteries, lithium batteries, and others. Batteries can also be classified by how they work, like primary batteries (single-use) and secondary batteries (rechargeable).

为了便于大家更好地认识电池王国的各大家族，接下来，我主要按照电池的能量转化方式以及工作性质的不同，来为大家介绍他们！

To help you better understand the different families in the Battery Kingdom, I will introduce them based on how they work and convert energy!

化学电池 Chemical Batteries	一次电池 Primary Batteries	干电池、碱锰电池、锂电池、锌－汞电池、镉－汞电池、锌－空气电池、锌－银电池、固体电解质电池（银－碘电池）等 Dry Battery, Alkaline Manganese Battery, Lithium Battery, Zinc-mercury Battery, Cadmium-mercury Battery, Zinc-air Battery, Zinc-silver Battery, Solid Electrolyte Battery (Silver-iodine Battery), etc.
	二次电池 Secondary Batteries	铅酸电池、镍镉电池、镍氢电池、锂离子电池、钠－硫电池等 Lead-acid Battery, Nickel-cadmium Battery, Nickel-metal Hydride Battery, Lithium-ion Battery, Sodium-sulfur Battery, etc.
	其他电池 Other Batteries	燃料电池、空气电池、纸电池、纳米电池等 Fuel Battery, Air Battery, Paper Battery, Nanobattery, etc.
物理电池 Physical Batteries	太阳电池 Solar Battery	

有一些电池是根据电极材料的区别来命名的。由于各自的制作材料和技术不一样，它们的"个性"也有很大不同。

Some batteries are named after the electrode materials. Because of the difference in materials and techniques, each battery has its own special "personality".

认识了这么多种电池,怎样判断一块电池的好坏呢?

Now that we know about so many types of batteries, how can we tell if a battery is good?

一块电池最基本的判断标准是**容量、功率、安全性**,需要通过各种测试来测验性能。

A battery's **capacity, power, and safety** are the basic things to check for, and these qualities can be tested through a variety of tests.

原来是这样,那你带我详细了解一下它们的性能吧!

Oh, I see! Then, let's take a closer look at their performance!

随着科技产品的更新迭代,人们对电池的要求也日益增多。**对环境的适应性、可移动性、是否适用于多种设备**等因素,都需要进行综合考量。

As technology moves forward, the demand for batteries is growing. We also need to think about **how well a battery adapts to different environments, how portable it is, and whether it can be used in many different devices.**

探秘锂离子电池
Exploring Lithium-ion Batteries

锂离子电池属于二次电池，它主要依靠锂离子在正极和负极之间移动来工作。

Lithium-ion batteries are a type of secondary battery that operate mainly by the movement of lithium ions between the anode and cathode.

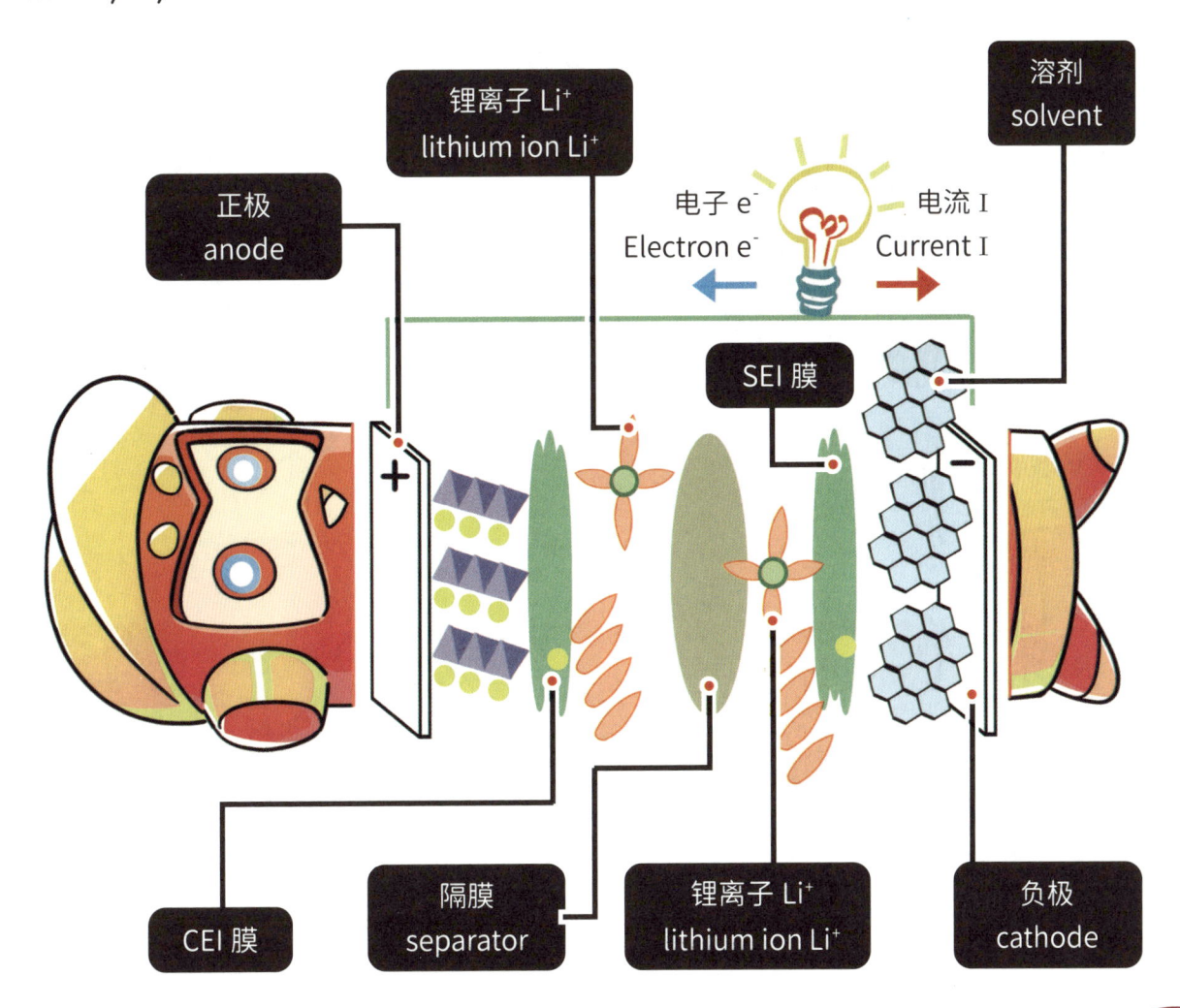

锂离子电池由于比能量大、重量轻、寿命长等优点，同时还是不含有毒物质的绿色电池，被人们广泛应用于轻便的可移动设备。

Due to the high energy density, lightweight nature, and long lifespan, lithium-ion batteries have become a popular choice for portable devices, as they are also environmentally friendly and free from toxic substances.

电池小知识 Battery tip

电池在一定条件下，对外做功所输出的电能，叫作"电池的能量"。"比能量"就是电池的单位质量，或单位体积所具有的有效电能量。

The "energy" of a battery refers to the electrical energy it outputs under certain conditions. "Specific energy" is the amount of effective electrical energy stored in the battery per unit mass or volume.

在便携式电子设备领域，随着手机、相机、笔记本电脑等设备向轻、薄、小方向发展，人们对电池的稳定性、连续使用时间、体积、充电次数和充电时间等的要求越来越高。

作为先进二次电池的代表，锂离子电池具备的质量轻、体积小、续航时间长等优点恰好满足这些要求，在便携式电子设备领域获得了绝对优势。

In the field of portable electronics, as devices like smartphones, cameras, and laptops continue to evolve towards being lighter, thinner, and more compact, there is an increasing demand for batteries with higher stability, longer usage time, smaller size, and faster charging.

As a representative of advanced secondary batteries, lithium-ion batteries boast advantages like light weight, small size, and long battery life, making them perfectly cater to these demands. This has granted them a dominant position in the portable electronics market.

在化学电池中，有多个电池家族的成员可作为可充电便携式电池。比如现在随处可见的"充电宝"，就是我们锂离子电池家族成员的工作领域之一。

In chemical batteries, there are several families suitable as rechargeable portable batteries. For example, the commonly seen "power bank" is one of the applications of the lithium-ion battery family.

锂离子电池
Lithium-ion Battery

也有别的电池家族在这个领域工作，例如铅酸蓄电池家族、镍镉电池家族、镍氢电池家族。

Other battery families, such as lead-acid, nickel-cadmium, and nickel-metal hydride, also serve in this field.

不过由于我们的能量密度最高、最轻巧，在充电及放电过程中的效率也较高，所以我们最受人们欢迎！

However, due to our highest energy density, lightweight design, and high efficiency during charging and discharging, we are the most popular choice!

目前，在人们的日常生活中，提高纯电动汽车的续航里程是最为迫切的需求。

锂离子电池的质量与比能量最高，是动力电池的首选，现在已经被广泛应用于混合动力电动汽车和纯电动汽车。

随着电动汽车的普及，锂离子电池产业迅速扩张。

In daily life, one of the most pressing needs is to increase the range of pure electric vehicles.

Lithium-ion batteries, with their high energy density and specific energy, are the top choice for power batteries. They are now widely used in both hybrid and pure electric vehicles.

As electric vehicles become more common, the lithium-ion battery industry is expanding rapidly.

不同类型的电动汽车，电池安装位置略有不同。

The placement of batteries in different types of electric vehicles vary slightly.

纯电动汽车的底盘比较宽敞，而且电池比较大，所以基本都被安装在底盘里。

Pure electric vehicles typically have spacious chassis and large batteries, making the chassis an ideal place for battery placement.

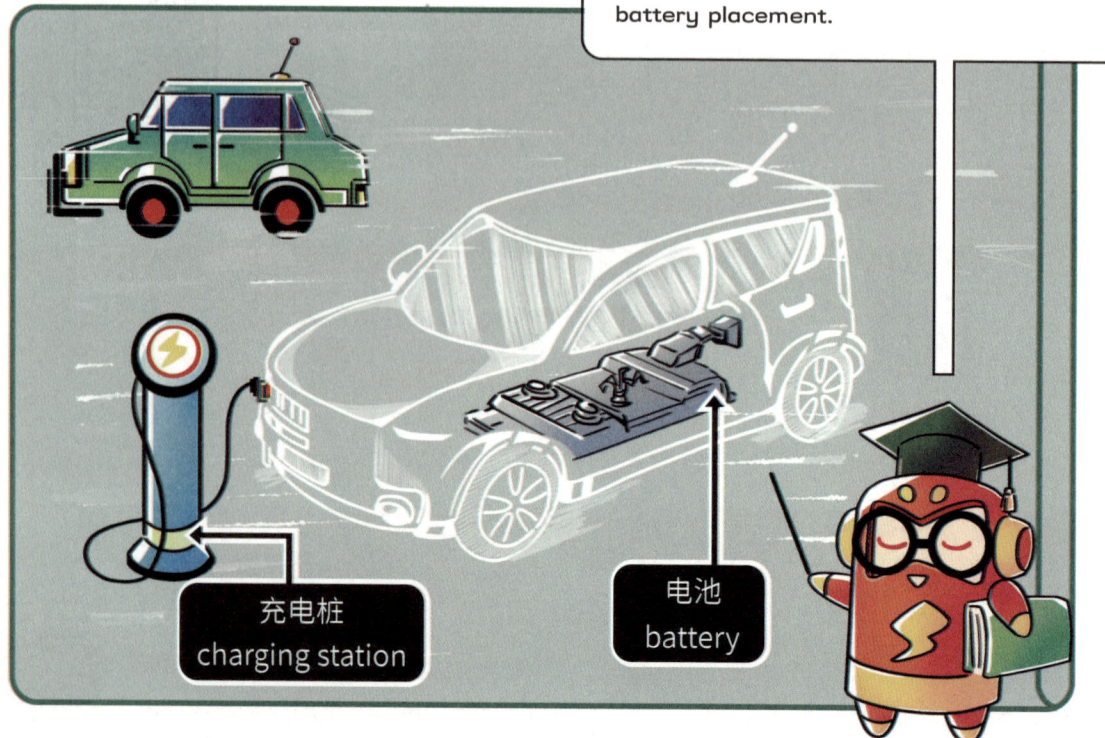

充电桩
charging station

电池
battery

锂离子电池的维护与保养
Maintenance and Care of Lithium-Ion Batteries

电池的寿命取决于反复充放电次数。充放电次数越多，电池寿命越短。

放电的深度是影响电池寿命的主要因素，放电的深度越高，电池的寿命就越短。换句话说，只要降低放电深度，就能大幅延长电池的使用寿命。大家在使用手机、笔记本电脑、照相机等便携式电器时，应及时充电，避免电量完全耗尽再充电。

The lifespan of a battery is determined by how many times it can be charged and discharged. The more often a battery is charged and used, the shorter its life will be.

One of the main factors that affects how long a battery's life lasts is the "depth of discharge". The deeper the battery is discharged, the shorter its life will be. In simple terms, the less you use up the battery's charge each time, the longer it will last. When you use devices like phones, laptops, or cameras, remember to charge them in time. It's better not to wait until the battery runs out before recharging.

我们喜欢"浅充浅放"，就是说：充电不充满，放电不放完。一般来说，保持在20%～80%的电量，能使我们的工作寿命更长哦！

We like to be "partial charging and discharging", which means: don't fully charge it, nor let it run out of power. So, here's a tip: try to keep your battery's charge between 20% and 80%. This helps your battery "live" longer!

大家千万记住，虽然我们锂离子电池的工作温度在 −20 ℃～ 60 ℃，但禁止暴晒，或者冰冻，尤其要避免将我们长时间"存放"在停止使用的设备中！

Finally, remember that while lithium-ion batteries work in temperatures from −20℃ to 60℃, we should never be exposed to direct sunlight or freezing cold. Also, make sure to avoid leaving us unused in devices for long periods.

探秘铅酸蓄电池
Exploring Lead-acid Batteries

迄今为止，铅酸蓄电池已经历了160多年的发展历程，可以说是电池大家族里的"老人"了！

Lead-acid batteries have been around for more than 160 years, making them the "elder people" of the battery family!

作为早期出现的电池，铅酸蓄电池工艺积累了大量的经验，因此这种电池在各方面都得到了长足的进步。

铅酸蓄电池
Lead-acid Battery

As one of the earliest types of battery, lead-acid battery technologies have gained a lot of experience over the years, growing in many ways.

铅酸蓄电池由于具备经济适用、维护简单、使用寿命长，以及质量稳定、可靠性高等优点，不论是在交通、通信、电力、军事，还是在航海、航空等领域，都发挥了重要作用。

These batteries are known for being cost-effective, easy to maintain, and long-lasting. They are also very stable and reliable, which is why they are used in many important areas, such as transportation, communication, electricity, the military, and even in fields like maritime and aviation.

铅酸蓄电池
lead-acid battery

电动摩托车
electric motorcycle

在我们的生活中，常见的电动自行车里面的"电瓶"，大多都采用铅酸蓄电池。

You've probably seen "storage batteries" in electric bicycles or electric motorcycle—those are often powered by lead-acid batteries!

电动自行车和电动摩托车等电动车已成为人们生活中的重要交通工具，它们的使用寿命是消费者们最为关心的问题之一。

Electric bicycles and electric motorcycle have become important transportation in our daily lives. The lifespan of these vehicles is one of the major concerns for users.

电池作为电动自行车和电动摩托车的核心部件之一，其使用寿命在很大程度上决定了这两种交通工具的使用寿命。因此，应加强对电池的保养，以便延长二者的使用寿命。

The battery, as one of the key components of electric bicycles and electric motorcycle, largely determines how long these vehicles will last. Therefore, taking good care of the battery can help extend their lifespan.

铅酸蓄电池的维护与保养
Maintenance and Care of Lead-acid Batteries

对于铅酸蓄电池的保养，首先要注意的就是"天天使用，天天充电"，否则就会出现"亏电"的情况。

When it comes to the maintenance of lead-acid batteries, the first thing to pay attention to is "use it every day, charge it every day", otherwise it would become "undercharged".

铅酸电池充电器
Lead-acid Battery Charger

电池最怕的就是"亏电"欠压，如果常常"亏电"，电池极板很容易受伤，这种损伤是无法修复的。所以，应保证电池随时有充足的电压。

Batteries fear being "undercharged" the most. If they are often undercharged, the battery plates can get damaged, and this damage can't be fixed. So, it's important to keep the battery well-charged.

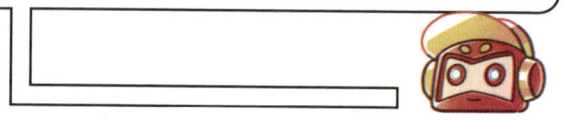

其次，影响电池寿命的因素，除了过度放电以外，还需要注意过度充电。如果充电时间过长，就要检查充电器电压保护装置有没有坏损。如果使用不适配或已损坏的充电器，则极易充坏电池。另外，如果长期使用快速充电器给电动自行车充电，同样对电池的极板有伤害。

Secondly, in addition to over-discharging, over-charging is another factor that affects battery life. If the charging time is too long, check whether the charger voltage protection device is damaged. If an unsuitable or damaged charger is used, the battery is easily damaged. In addition, if a fast charger is used to charge an electric bike for a long time, it will also damage the battery plates.

探秘碱锰电池
Exploring Alkaline-manganese Batteries

一次电池即用完即弃的电池，因为它们的电量耗尽后，无法再充电和使用，只能被丢弃。

最常见的一次电池就包括碱锰电池。最常见的碱锰电池有圆筒形和纽扣形两种，此外还有方形和扁形等品种。

Primary batteries, commonly referred to as "single-use" batteries, cannot be recharged and use again once their energy is depleted and are disposed of after use.

Among the most widely used primary batteries are alkaline-manganese batteries. These are typically available in cylindrical and button shapes, and other forms like rectangular and flat shapes also exist.

碱锰电池
alkaline-manganese batteries

碱锰电池是以锌为负极，二氧化锰为正极，氢氧化钾溶液为电解液的一次电池。由于氢氧化钾溶液的凝固点较低、内阻小，因此，碱锰电池能在低温环境下工作，并仍然能大电流放电。

优质的碱锰电池，保存期可在5年以上，这是其他一次电池难以达到的。碱锰电池因具备使用方便、性能优良、对环境的适应性强以及贮存期长等优点，特别受到军事装备的青睐。在野战条件下，没有市电，充电又极困难，全靠电池供电，贮备的碱锰电池可以说是通信装备的"粮食"！

An alkaline manganese battery is a primary battery with zinc as the negative electrode, manganese dioxide as the positive electrode and potassium hydroxide solution as the electrolyte. Thanks to the low freezing point and low internal resistance of that solution, these batteries can function effectively in low-temperature environments and deliver high-current output.

High-quality alkaline-manganese batteries have a shelf life of over five years, a feat that is hard to achieve for other primary batteries. These batteries are particularly favored in military applications due to their ease of use, excellent performance, strong adaptability to various environments, and long shelf life. In field operations, where there is no access to electricity and recharging is nearly impossible, batteries become the sole power source. In such a case, stored alkaline-manganese batteries serve as the "lifeline" of communication equipment.

军用：战术电台、野战电话、仪器仪表、储备电池……
Military use: tactical radios, field telephones, measuring instruments, backup batteries, etc.

民用：照相机、对讲机……
Civilian use: cameras, walkie-talkies, etc.

瞧瞧，这些都是我们工作的地方，不敢小瞧我们了吧！
Check it out—these are just a few of the places we work. Bet you won't underestimate us now!

我们碱锰电池家族虽然属于一次电池，不过我们的容量大，保存时间长，应用范围广。

As primary batteries, we stand out with our large capacity, long shelf life, and wide range of applications.

可以用作遥控器、寻呼机、测试仪表、收音机、手持对讲机等装置的配套电源。

We serve as reliable power sources for devices such as remote controls, pagers, testing instruments, radios, and handheld walkie-talkies.

为什么她穿着羽绒服？

Why is she wearing a down jacket?

在气温较低地区的户外使用照相机，尤其要注意相机或电池的保暖。

In colder regions, extra care must be taken to keep cameras or batteries warm during outdoor use.

由于数码相机中的电池在气温过低的环境下，其活性物质的活跃度大大降低，因而可能无法提供相机正常工作的电流。

This is because, at low temperatures, the active materials in batteries used in digital cameras become significantly less active, potentially failing to provide the necessary operating current for the camera.

碱锰电池家族很优秀，小小的身体有大大的能量，能在高负荷下连续工作的同时维持较高的稳定电压，并且在 –20 ℃～ 60 ℃之间都可以稳定工作。

Despite thier small size, they carry immense energy! Thier family excels in maintaining a stable, high voltage even under heavy loads. Moreover, they can operate reliably across a wide temperature range, from –20 ℃ to 60 ℃ .

曾经的碱锰电池被称为"碱性锌锰电池",当时它的内部还含有对人体有害的元素——汞。随着环保电池的普及,含汞电池逐渐被淘汰。

由于碱锰电池具有优异的性能,科学家们对它们进行了无汞改良,改良后的产品被称为"无汞碱锰电池"。

目前,我国在碱锰电池的无汞化技术研究方面,仍受装备、零配件和材料等因素的制约,还有很大的发展潜力。

In the past, alkaline-manganese batteries were known as "alkaline zinc-manganese batteries". At that time, they contained mercury, a toxic element harmful to human health. However, as environmentally friendly batteries gained popularity, mercury-containing batteries were gradually phased out.

Thanks to the excellent performance of alkaline-manganese batteries, scientists worked to develop mercury-free versions of these batteries. These improved batteries are now known as "mercury-free alkaline-manganese batteries".

Currently, research into mercury-free technology for alkaline-manganese batteries in China faces constraints in areas such as equipment, components, and materials. Nevertheless, the field holds significant potential for further advancement.

探秘镍氢电池
Exploring Nickel-Metal Hydride Batteries

由于化石燃料在人类大规模开发利用的情况下越来越少，近年来，氢能源的开发利用日益受到重视。镍氢电池作为氢能源应用的一个重要方向越来越受到人们注意。

As fossil fuel reserves dwindle due to large-scale human exploitation, the use of hydrogen has become increasingly prominent in recent years. As a key application of hydrogen energy, nickel-metal hydride (NiMH) batteries are also gaining significant attention.

镍氢电池
NiMH batteries

镍氢电池是一种性能良好的电池，被称作"最环保"的电池，其回收再利用率比锂离子电池还要高，被誉为"人类最理想的终极能源"。

Renowned for their excellent performance, NiMH batteries are often referred to as the "most eco-friendly" batteries. With recycling efficiency surpassing that of lithium-ion batteries, they are hailed as "the ultimate energy source for humanity".

镍氢电池的特性与应用
Characteristics and Applications of NiMH Batteries

最初，镍氢电池仅仅只有高压镍氢电池这一种，但因其高成本、低安全性的缺点，只被应用在航天领域，并没有被广泛应用。

后来，科学家们研发出了低压镍氢电池，它们这才逐渐进入大众的视野，如在移动电话、无绳电话、便携式照相机、笔记本、应急灯、家用电器等设备中，都出现了镍氢电池的身影。

Initially, NiMH batteries were available only as high-voltage versions. However, due to their high cost and low safety, their use was limited to aerospace applications and was not widely used.

Later, scientists developed low-pressure nickel-metal hydride batteries, which gradually gained public attention. They began to appear in various devices such as mobile phones, cordless phones, portable cameras, laptops, emergency lights, and household appliances.

镍氢电池家族的可靠性强，循环使用寿命长，可达到数千次之多，即使温度达到零下10度，其性能都不受影响，而且可以快速充放电，是各种现代设备的理想之选。

The NiMH battery family boasts exceptional reliability and a remarkably long cycle life, capable of enduring thousands of charge-discharge cycles. Even at temperatures as low as −10 ℃, their performance remains unaffected. Additionally, they support rapid charging and discharging, making them an ideal choice for various modern applications.

目前，镍氢电池除了被广泛应用于移动通讯、笔记本计算机等各种小型便携式的电子设备以外，更大容量的镍氢电池已经开始被用于汽油/电动混合动力汽车，世界各国都在加紧此项技术的研究。

汽车在低速行驶状态时，通常会比在高速行驶状态下消耗更多的汽油。利用镍氢电池可快速充放电的特性，在汽车高速行驶时，发电机所发的电可储存在车载的镍氢电池中；在低速行驶时，利用车载的镍氢电池驱动电动机来代替内燃机工作。这样，既可以保证汽车正常行驶，又节省了汽油。

Beyond their widespread use in portable electronic devices like mobile phones and laptops, higher-capacity NiMH batteries are now being adopted for gasoline and electric hybrid vehicles. Countries worldwide are intensifying their efforts in research to enhance this technology.

Cars tend to consume more fuel at low speeds compared to high speeds. NiMH batteries, with their rapid charging and discharging capabilities, offer an efficient solution: during high-speed driving, the electricity generated by the car's alternator is stored in the on-board NiMH battery; during low-speed driving, the stored energy powers the electric motor, replacing the internal combustion engine. This system ensures smooth vehicle operation while reducing gasoline consumption, making hybrid vehicles more fuel-efficient.

电池小知识 Battery tip

镍氢电池家族离不开一种稀土材料。将稀土材料应用于镍氢电池的研制与开发，能够起到改善电池的性能以及延长使用其寿命的作用。我国拥有丰富的稀土金属资源，这是我国能源领域的一大骄傲！

The NiMH battery family owes much of its success to a special material: rare earth elements. Integrating rare earth materials into NiMH battery development improves the performance and extends the life span. China possesses abundant rare earth metal resources, which is a significant source of pride for the country's energy sector!

探秘镍镉电池
Exploring Nickel-Cadmium Batteries

镍镉电池是最早被应用于手机的电池种类，具有良好的大电流放电特性，耐过充和耐过放能力强，维护简单，经济耐用，是一种理想的直流供电电池。

Nickel-cadmium (NiCd) batteries were the first type of batteries used in mobile phones. They are known for their excellent high-current discharge characteristic and strong resistance to over-charging and over-discharging. They are easy to maintain, affordable and very durable, making them an ideal choice for direct current (DC) power supply.

镍镉电池
nickel-cadmium batteries

我们的寿命长，放电性能优异，工作温度适应性强，在零下 30 ℃～50 ℃之间都没问题。

We have a long lifespan and great discharge performance. We can operate in a wide range of temperatures, from −30 ℃ to 50 ℃.

我们的工作领域还很广呢！任何无线设备的相关工作，我们都有能力胜任！

We have a wide range of applications! We can support any task related to wireless devices.

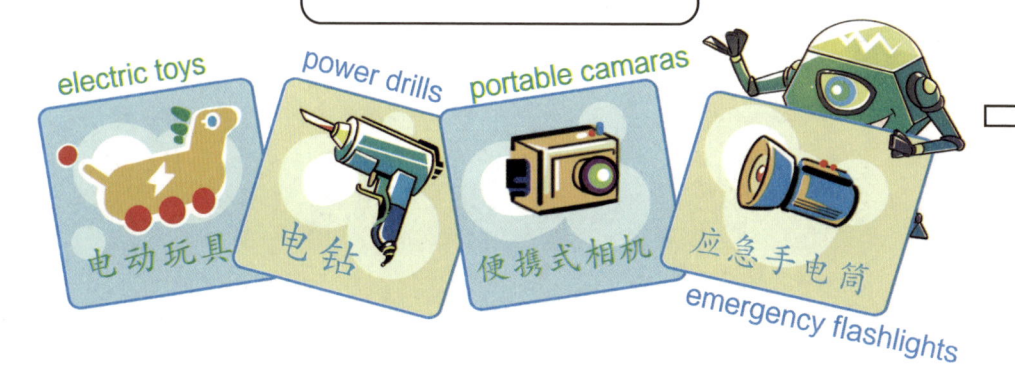

electric toys
电动玩具

power drills
电钻

portable camaras
便携式相机

emergency flashlights
应急手电筒

因为含有"镉"这种对人体有害的重金属，一旦镍镉电池报废之后，就需要专门回收，以免污染环境。

Once NiCd batteries are scrapped, they need to be recycled with special care as they contain cadmium that is detrimental to humans and could pollute the environment if tossed around.

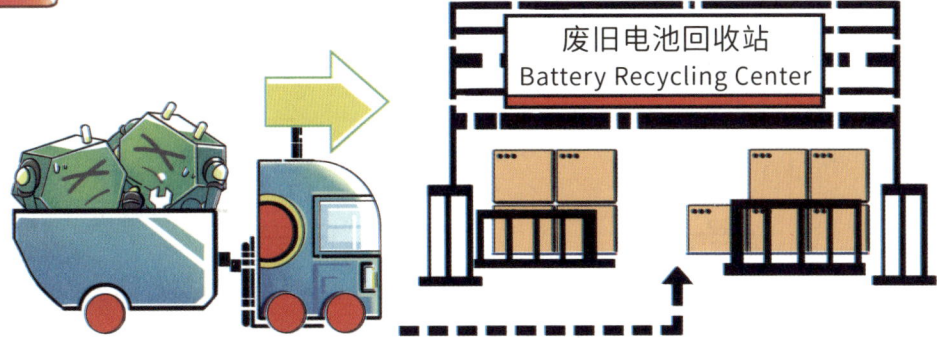

镍镉电池体内含有大量的镍、镉和铁，其中镍是钢铁制造、电器生产、有色合金冶炼以及电镀工业等多个领域的重要原料，镉是电池制造、颜料生产和合金冶炼等多个领域使用的稀有金属，但这种金属有剧毒，如果将废弃电池乱扔的话，会污染土壤、水域等环境。食用在这些环境中生长的蔬菜、鱼类等食品，会对人的健康造成伤害。

现在，镍镉电池因对人体健康和生态环境危害较大，已被列入《国家危险废物名录》。

NiCd batteries contain huge amounts of nickel, cadmium, and iron. Nickel is an essential raw material in steel manufacturing, electrical production, non-ferrous alloy smelting, and the electroplating industry; cadmium is a rare metal used in battery manufacturing, pigment production, and alloy smelting, but it is highly toxic. If scrapped batteries are not disposed of properly, they can pollute the soil and water. Vegetables, fish, and other foods grown in these polluted environments are harmful to human health.

Nowadays, due to their significant harm to human health and the ecological environment, NiCd batteries have been listed in the *National Catalogue of Hazardous Wastes*.

再告诉你一个小秘密，镍镉电池很"记仇"，在充放电过程中如果处理不当，会出现严重的"记忆效应"，这样就会使它的寿命大大缩短。

Tell you a secret: NiCd batteries are very 'resentful'. If not charged and discharged properly, they can develop a serious 'memory effect' that greatly shortens their lifespan.

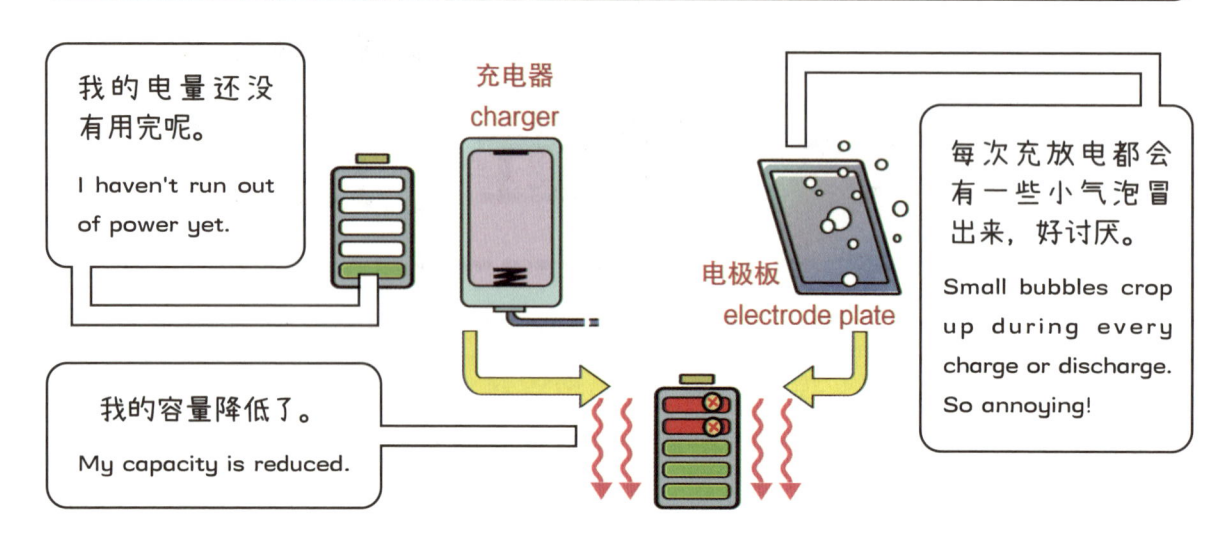

我的电量还没有用完呢。
I haven't run out of power yet.

充电器
charger

电极板
electrode plate

每次充放电都会有一些小气泡冒出来，好讨厌。
Small bubbles crop up during every charge or discharge. So annoying!

我的容量降低了。
My capacity is reduced.

"记忆效应"又称"电池结晶效应"，指电池在充电前，电池的电量没有被完全放尽，久而久之，将会引起电池容量的降低。在电池充放电的过程中，电池极板上会产生些许的小气泡，日积月累，这些气泡减少了电池极板的有效面积，也间接影响了电池的容量。

不过这种效应一般只会发生在镍镉电池身上，可以通过掌握合理的充放电方法来减轻"记忆效应"。

The "memory effect", or the "battery crystallization effect", happens when the batteries are recharged without being fully discharged, thus leading to a decrease in battery capacity. During the charge-discharge cycles, small bubbles appear on the electrode plates. These bubbles gradually reduce the surface active area of the plates and indirectly affect the battery capacity.

Generally, this effect only occurs in NiCd batteries and can be reduced by proper charging and discharging methods.

探秘镍镉电池与镍氢电池
Exploring Nickel-cadmium Batteries and Nickel-metal Batteries

镍氢电池是由镍镉电池改良而来的，其正极处的化学反应类似于镍镉电池。它们的正极都是使用镍氢氧化物。但是，镍氢电池的负极是以能吸收氢的金属代替镉。

NiMH batteries are improved versions of NiCd batteries. They have similar chemical reactions at the positive electrode, which both use nickel hydroxide, but NiMH batteries use hydrogen-absorbing metals instead of cadmium at the negative electrode.

咦？那里好像有别的电池人！

Huh? It seems to be some other battery there!

那应该是镍镉电池家族的人。自从人类开始用镍氢电池代替镍镉电池以后，他们就面临着即将失业的风险，行为变得有些鬼鬼祟祟。

That must be the NiCd battery family. Since people started using nickel-metal hydride (NiMH) batteries to replace them, they've been at risk of unemployment and have become a bit sneaky.

镍氢电池与镍镉电池相比，有三个最主要的优势：

第一，能量密度比较高。与相同体积镍镉电池对比，镍氢电池容量是镍镉电池的两倍，这意味着在不额外增加用电设备重量时，使用镍氢电池能大大延长设备工作时间。

第二，大大减少了镍镉电池中存在的"记忆效应"问题，从而可更方便地使用。

第三，比镍镉电池更环保。因为它内部没有有毒重金属镉元素。

作为二次电池家族的后起之秀，镍氢电池以相同的价格提供比镍镉电池更高的电容量，具有不明显的"记忆效应"，以及比较低的环境污染（不含有毒的镉），逐渐开始在各个应用领域替代镍镉电池的工作。

NiMH batteries have three main advantages over NiCd batteries:

Firstly, higher energy density: NiMH batteries have twice the capacity of NiCd batteries of the same size. This allows devices to operate significantly longer without adding extra weight.

Secondly, reduced "memory effect": NiMH batteries greatly reduced the "memory effect" found in NiCd batteries, making them easier to use.

Thirdly, environmental friendliness: NiMH batteries are more environmentally friendly than NiCd batteries, because they do not contain the toxic heavy metal cadmium.

As a newcomer to the secondary battery family, NiMH batteries offer higher capacity than NiCd batteries with the same price, and have a minimal "memory effect", also cause less environmental pollution since they do not contain toxic cadmium. They are gradually replacing NiCd batteries in various applications.

探秘锌 - 空气电池
Exploring Zinc-air Batteries

锌－空气电池是一种体积小、电荷容量大、质量小、能在宽广的温度范围内正常工作、无腐蚀且安全可靠的环保电池。

Zinc-air batteries are small and light batteries that are good for the environment. They have a large charge capacity and can work safely in many different temperatures. These batteries do not rust and are very reliable.

锌－空气电池以空气中的氧气为正极活性物质，金属锌为负极活性物质，以氯化铵或苛性碱溶液为电解质，具有安全、零污染、高能量、大功率、低成本及材料可再生等优点。

Zinc-air batteries use oxygen from the air as the positive electrode active material, zinc as the negative electrode active material, and ammonium chloride or caustic solution as the electrolyte. They have advantages such as high safety, zero pollution, high energy density, high power output, low cost, and material renewability.

纽扣形
锌－空气电池

button-shaped zinc-air battery

你们看，由我驱动的电动自行车连续行驶里程可以达 200 千米。

Look, the electric bicycle powered by me can travel over 200 kilometers continuously.

大型的锌－空气电池主要用于铁路和航海灯标装置上。纽扣形锌－空气电池已被广泛用于助听器中。

Large zinc-air batteries are used in railway devices and marine beacon devices. Button-shaped ones are widely used in hearing aids.

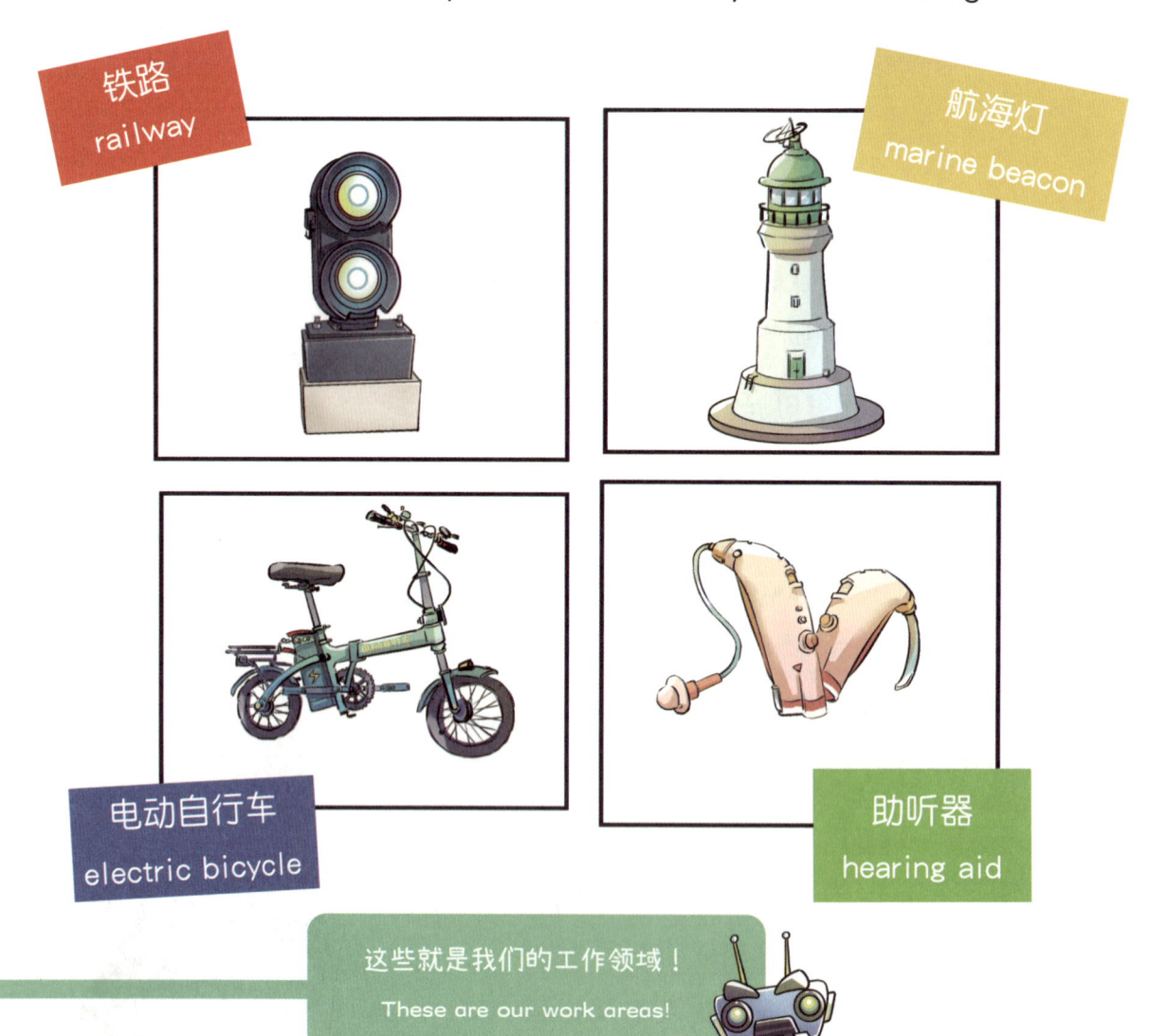

铁路
railway

航海灯
marine beacon

电动自行车
electric bicycle

助听器
hearing aid

这些就是我们的工作领域！
These are our work areas!

虽然现在批量生产锌－空气电池的工艺还不成熟，不过电池王国在这一领域正在不断发展，未来，他们有可能成为电动汽车的理想动力电源呢！

Although zinc-air batteries are not yet mature in its mass production, the Battery Kingdom is continuously developing in this field. In the future, they might become the ideal power source for electric cars!

探秘钠－氯化镍电池
Exploring Sodium-nickel Chloride Batteries

钠－氯化镍电池也被称为"钠盐电池",是在钠－硫电池研制基础上发展起来的一种新型高能热电池。

Sodium-nickel chloride batteries, also known as "sodium salt batteries", are a new type of high-energy thermal battery, which is developed based on the research of sodium-sulfur batteries.

告诉你一个电池王国的小秘密,王一氯和我们的国王王一硫,有着非同一般的关系,可以说是兄弟!

Here's a secret of the Battery Kingdom: King Monochlorine and King Monosulfide have an extraordinary relationship, just like brothers!

怪不得名字这么像……

No wonder their names are similar…

钠-氯化镍电池具有性质稳定、安全性高、使用寿命长、应用范围广、原材料易获得并且无毒、废品回收工艺简单且无污染等特点。

These batteries are stable, safe, and long-lasting. They can be used in many ways, and are made from easy-to-find, non-toxic materials. In addition, they can be recycled easily and without pollution.

钠-氯化镍电池
sodium-nickel chloride battery

钠-氯化镍电池家族成员的工作温度比较高，在 250～350 ℃范围内，由于他们自带加温设置，所以工作时外在温度对他们影响不大，即使进行 100 次冷热循环，他们的容量和寿命都没有丝毫影响。

Sodium-nickel chloride batteries work at high temperatures, between 250 ℃ and 350 ℃. They have a special heating system, so outside temperatures don't affect them much when working. Even after 100 hot and cold cycles, they still keep their power and last a long time.

钠-氯化镍电池又被称为"Zebra 电池"，如果你跟他们打招呼，叫他们斑马电池，他们也会答应的。

Sodium-nickel chloride batteries are also known as "zebra batteries". If you greet them by calling them zebra batteries, they will respond too.

它们和斑马是亲戚吗？　Are they related to zebras?

哈哈！当然不是。这里的"Zebra"，其实是英文"zero emission battery research activity"的缩写，表明其是一种零排放无污染的绿色电源。

Haha! Of course not. The "Zebra" here actually stands for "Zero Emission Battery Research Activity", meaning a zero-emission, pollution-free green power source.

钠-氯化镍电池最突出的优势就在于它的安全性好、可靠性高，具有过热状态下不会着火、爆炸等特性，并且电池性能与周围环境温度完全无关，能在恶劣环境下工作。

The best thing about sodium-nickel chloride batteries is that they are very safe and reliable. They won't catch fire or explode, even if they get too hot. Their performance doesn't change with the temperature around them, so they can work in tough conditions.

我在工作时性能稳定，通过了多种安全性与可靠性测试，即使比赛发生再大的意外，我也能镇定自若救场，保障比赛的安全性。

I work stably and have passed many safety tests. Even if a major accident happens during a competition, I can keep calm and handle the situation to ensure the safety of the event.

钠-氯化镍电池通过了极为严格试验的安全考核试验，甚至可以承受如海水浸泡、子弹冲击、高空坠地、燃烧，以及极寒极热等极端条件。

虽然性能优异，但钠-氯化镍电池在工作时，需要保持在300 ℃左右的温度，热循环启动需要12～15小时，目前主要被应用于电网储能、通讯基站及动力电源领域。

Sodium-nickel chloride batteries have passed very strict safety tests. They can endure extreme situations like being soaked in seawater, being hit by bullets, falling from high places, and burning. Besides, they can work in extreme conditions such as extreme cold and extreme heat.

Even though they work really well, sodium-nickel chloride batteries need to maintain about 300 ℃ during operation. It takes 12 to 15 hours to get them ready for use. Right now, they are mainly used in grid energy storage, communication base stations, and power supply fields.

探秘钠－硫电池
Exploring Sodium-sulfur Batteries

> 我们来拜访国王。
> We come here for the King.

> 他去视察风力发电厂了。
> He is inspecting the wind power plant.

　　钠－硫电池作为一种新型化学电源，自问世以来已有了很大发展。钠－硫电池体积小、容量大、寿命长、效率高，被广泛应用于削峰填谷、应急电源、风力发电等电力储能方面。

Sodium-sulfur batteries are a new type of chemical power source, and they have grown a lot since they were first created. With their small size and large capacity, these batteries can last a long time and are very efficient. They are widely used in energy storage for things like balancing electricity supply, emergency power, and wind power generation.

钠－硫电池由熔融电极和固体电解质组成，正极活性物质为液态硫和多硫化钠熔盐，负极活性物质为熔融金属钠。

Sodium-sulfur batteries are made of liquid electrodes and solid electrolytes. The positive electrode is made of liquid sulfur and sodium polysulfide molten salt, while the negative electrode is made of molten sodium metal.

钠－硫电池
sodium-sulfur batteries

太阳能、风能等新能源虽然洁净，但发电功率很不稳定。这会给整个电网带来不期而至的"洪峰"。储能电站会将这些"绿电"先照单全收，再根据电网需求输出。

它的"蓄洪"性能非常优异，即使输入的电流突然超过额定功率的5～10倍，它也能泰然承受，再以稳定的功率释放到电网中——这对于大型城市电网的平稳运行尤其重要。

New energy sources like solar energy and wind energy are clean, but their power is not stable. This can cause sudden "power surges" in the whole power grid. Energy storage stations can first store this "green energy" and then release it when needed, helping balance the power supply according to the grid's demands.

Sodium-sulfur batteries are very good at "storing" power. Even if the incoming current suddenly becomes 5 to 10 times stronger than the normal power, the battery can handle it calmly and release power steadily into the grid. This is especially useful for keeping power grids running smoothly in large cities.

这是钠-硫电池家族工作最重要的领域之一——为电站负荷调平。

One of the most important jobs of sodium-sulfur batteries is to help balance the power supply at stations.

将夜晚多余的电存储在电池里,到白天用电高峰时将其从电池中释放出来。

They store extra electricity at night and then release it at peak hours of the day.

有一个专有名词来解释这种作用,那就是"削峰填谷"。

This process is called "peak shaving and valley filling".

钠-硫电池具有独到的储能优势,主要体现在原材料和制备成本低、能量和功率密度大、效率高、不受场地限制、维护方便等方面。

它的工作原理是——钠与硫通过化学反应,将电能转化为化学能储存,当电网需要更多电能时,它又会将化学能转化成电能,释放出去。

Sodium-sulfur batteries excel in storing energy. They are efficient and cheap, easy to maintain, and have high energy, power density, and can be used in many places.

How do they work? Sodium and sulfur react chemically to store electrical energy as chemical one. When the grid needs more power, they turn the chemical energy back into electricity and release it.

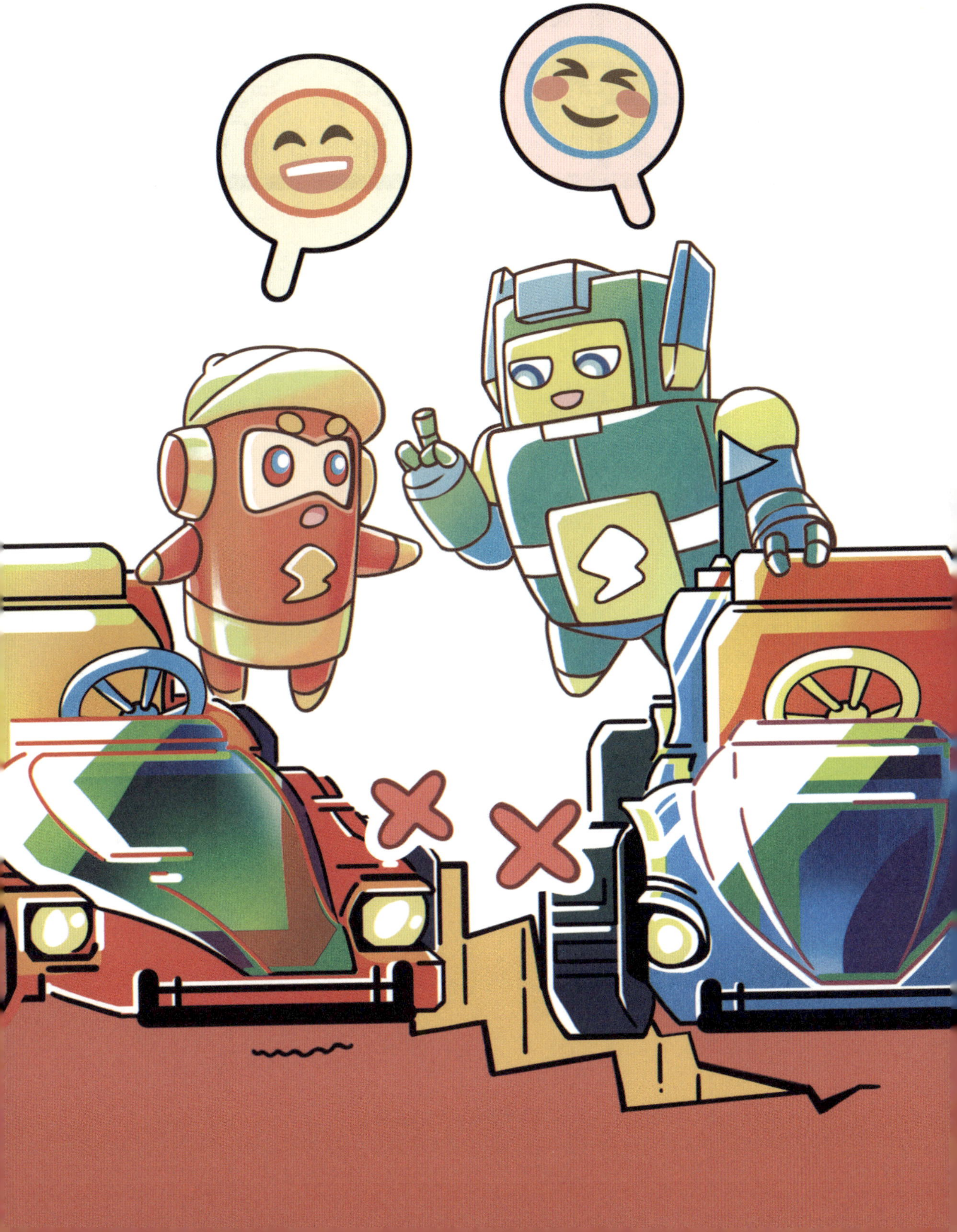

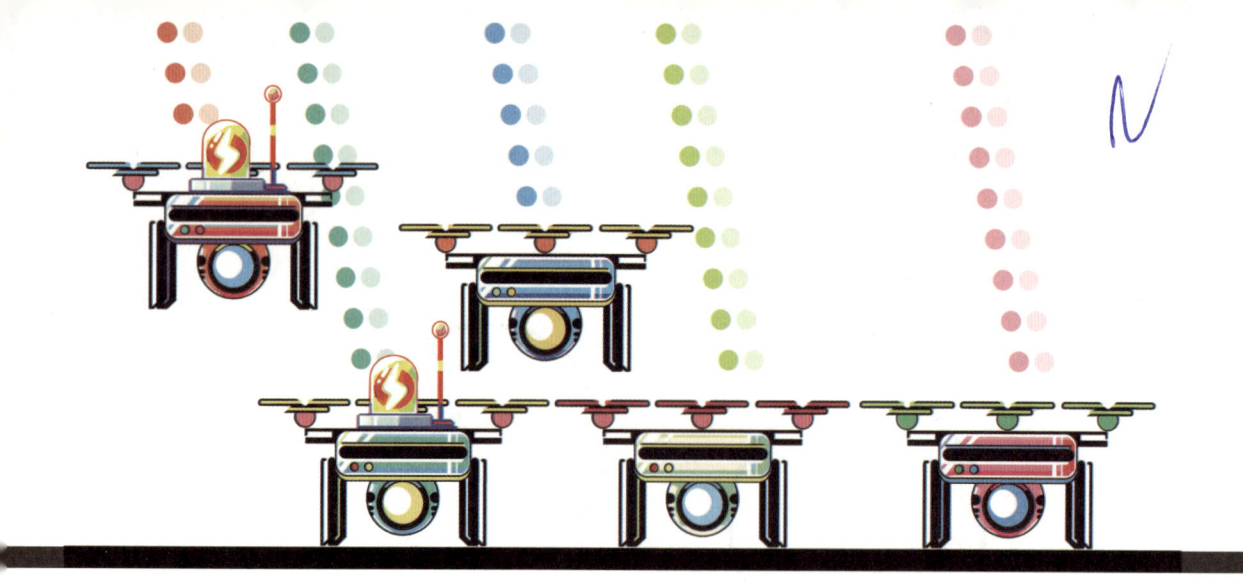

图书在版编目（CIP）数据

电池大家族：汉英对照 / 马建民主编；咪柯文化
绘图；何敏译. -- 成都：成都电子科大出版社，2025.
1. -- ISBN 978-7-5770-1486-9

Ⅰ.TM911-49

中国国家版本馆 CIP 数据核字第 2025H8L356 号

电池大家族（中英对照版）
DIANCHI DA JIAZU（ZHONG-YING DUIZHAO BAN）

马建民　主编　咪柯文化　绘　何　敏　译

策划编辑	谢忠明　段　勇
责任编辑	赵倩莹
责任校对	蒋　伊
责任印制	段晓静

出版发行	电子科技大学出版社
	成都市一环路东一段 159 号电子信息产业大厦九楼　邮编 610051
主　　页	www.uestcp.com.cn
服务电话	028-83203399
邮购电话	028-83201495

印　　刷	成都久之印刷有限公司
成品尺寸	185 mm×260 mm
印　　张	9
字　　数	172 千字
版　　次	2025 年 1 月第 1 版
印　　次	2025 年 1 月第 1 次印刷
书　　号	ISBN 978-7-5770-1486-9
定　　价	66.00 元

版权所有，侵权必究